Reliability a

K. N. Shukla
Akash Kumar Shukla
Geetam Richhariya

Reliability and Availability Evaluation of Hydro Power System

LAP LAMBERT Academic Publishing

Imprint

Any brand names and product names mentioned in this book are subject to trademark, brand or patent protection and are trademarks or registered trademarks of their respective holders. The use of brand names, product names, common names, trade names, product descriptions etc. even without a particular marking in this work is in no way to be construed to mean that such names may be regarded as unrestricted in respect of trademark and brand protection legislation and could thus be used by anyone.

Cover image: www.ingimage.com

Publisher:
LAP LAMBERT Academic Publishing
is a trademark of
International Book Market Service Ltd., member of OmniScriptum Publishing Group
17 Meldrum Street, Beau Bassin 71504, Mauritius
Printed at: see last page
ISBN: 978-620-2-55561-6

Copyright © K. N. Shukla, Akash Kumar Shukla, Geetam Richhariya
Copyright © 2020 International Book Market Service Ltd., member of OmniScriptum Publishing Group

Book on

Reliability and Availability Evaluation of Hydro Power System

Author (s)

Dr. K.N.Shukla,
Professor, Department of Electrical and Electronics Engineering, LNCT Bhopal

Dr. Akash Kumar Shukla
Assistant Professor, Department of Electrical Engineering, UEC Ujjain

Dr. Geetam Richhariya
Professor, Department of Electrical Engineering, NRI Group of institution, Bhopal

PREFACE

The importance of reliability and availability evaluation of hydro power system power system reliability is demonstrated when our electricity supply is disrupted, whether it decreases the comfort of our free time at home or causes the shutdown of our companies and results in huge economic deficits. The objective of Reliability and Availability Evaluation of Hydro Power System is to contribute to the improvement of hydro power system reliability. *Reliability and Availability Evaluation of Hydro Power System* has been written in straightforward language that continues into the mathematical representation of the methods. Power engineers and developers will appreciate the emphasis on practical usage, while researchers and advanced students will benefit from the simple examples that can facilitate their understanding of the theory behind hydro power system reliability and that outline the procedure for application of the presented methods.

Author(s)

Dedicated to

Mata Matushree

INDEX

List of Figures..

List of tables..

Abbreviations ....

List of symbols...

Chapter 1 Introduction   11

    General..

    Organization of the book ....

Chapter 2 Fundamental of Reliability   14

    Reliability ....

    Power System Reliability ....

    Probabilistic reliability criteria ....

    Statistical and probabilistic measures..

    Absolute and relative measures ....

    Methods of assessment ....

    Basic definition related to reliability ....

    Failure ....

    Failure Density Function ....

Forced outage ....

Scheduled outage..

Maintenance, Maintainability...

    Availability and Unavailability ....

    Mean Time to Failure, Mean Time between Failures and Mean Time to Repair ...

Failure Rate and Repair Rate...

    Cycle Time and Cycle Frequency ....

Reliability Block Diagram ....

System Reliability Models ....

Non-Series- Parallel System..

Chapter 3 Markov Modeling and Estimation   34

Introduction ....

Continuous Markov Process..

Transition rate concepts...

Evaluating time dependent probabilities ....

Evaluating limiting state probabilities...

State Space Diagrams ....

Single repairable component ....

Two repairable components ....

Large number of components..

Standby redundant system...

Mission oriented system..

Multi–state problems ....

Two component repairable system ....
State Probabilities..
Frequency of encountering individual states ....
Cycle time between individual states ....
Discrete Markov Process ....
State Transition Probability Matrix ....
   The *n*-Step State Transition Probability ....

Chapter 4 Methodology and Mathematical Background   50

Methodology...

Mathematical Background..
Evaluating limiting state probabilities ....
Single repairable component ....
Two identical repairable components...
Evaluating time dependent state probability..
Unit Modeling...
Three-State Model ....

Chapter 5 Plant Modeling and Estimation for PHPS   60

Introduction ....
Unit Modelling ....
Plant Modelling ....
State Space Diagram...
State Probability, Availability and Reliability Determination............................

Chapter 6 Plant Modeling For CHPS   66

Introduction ....

Unit Modelling ....

Plant Modelling ....

State Space Diagram...

State Probability, Availability and Reliability Determination..............................

Chapter 7 Results and Discussion   73

Unit wise Reliability and Availability ....

Reliability and Availability from Plant Modelling ....

(a) For PATHRI Hydropower station ....

(b) For CHILLA Hydropower station..

Discussion and Conclusion..

Appendix 1.1..

Appendix 1.2..

Appendix 1.3..

Appendix 1.4..

Appendix 2.1..

Appendix 2.2..

Appendix 2.3..

References... 97

LIST OF FIGURE

Figure No.	Title
2.1	Components of system reliability
2.2	MTBF V_s MTTF
2.3	Two series component
2.4	Two component parallel system
2.5	Series – parallel system
2.6	Non-Series- Parallel system
3.1	Single component repairable system-State Space Diagram
3.2	State Space diagram of component with partial output state
3.3	State space diagram for two component system
3.4	State space diagram for two component standby system
3.5	State Space diagram for two identical non-repairable components
3.6	State space diagram for two component system
4.1	Two State model
4.2	Three state model
4.3	Developed hydro-unit model
5.1	State space diagram for PHPS
6.1	State space diagram for CHPS
7.1	Units Availability & Reliability histogram for PHPS 2005-2010
7.2	Units Availability and Reliability curves for PHPS 2005-2010
7.3	Units Availability & Reliability histogram for CHPS 2005-2010
7.4	Units Availability and Reliability curves for 2005-2010

LIST OF TABLE

Table No.	Title
3.1	The Rate of Departure or Entry
3.2	State Probabilities and Frequencies
4.1	State Probability Value
4.2	Frequency of Encountering States
5.1	PHPS [Unit-1] Failure Rates, Repair Rates State Probability and Availability, Reliability Determination 2005-10
5.2	PHPS [Unit-2] Failure Rates, Repair Rates State Probability and Availability, Reliability Determination 2005-10
5.3	PHPS [Unit-3] Failure Rates, Repair Rates State Probability and Availability, Reliability Determination 2005-10
5.4	PHPS State Probability and Availability, Reliability Determination 2005-10
6.1	CHPS [Unit-1] Failure Rates, Repair Rates State Probability and Availability, Reliability Determination 2005-10
6.2	CHPS [Unit-2] Failure Rates, Repair Rates State Probability and Availability, Reliability Determination 2005-10
6.3	CHPS [Unit-3] Failure Rates, Repair Rates State Probability and Availability, Reliability Determination 2005-10
6.4	CHPS [Unit-4] Failure Rates, Repair Rates State Probability and Availability, Reliability Determination 2005-10
6.5	CHPS State Probabilities and Availability, Reliability Determination 2005-10
7.1	System Availability and Reliability of PHPS 2005-10
7.2	System Availability and Reliability of CHPS 2005-10
7.3	Unit Major Faults That Affect the Reliability Indices for PHPS
7.4	Unit Major Faults That Affect the Reliability Indices for CHPS

ABBREVITATIONS

CHPS	:	Chilla Hydro Power Station
PHPS	:	Pathri Hydro Power Station
MTTR	:	Mean Time To Repair
MTBF	:	Mean Time Between Failure
MTTF	:	Mean Time To Failure

LIST OF SYMBOLS

λ	:	failure rate
μ	:	repair rate
m	:	mean time to failure
r	:	mean time to repair
A	:	availability
U	:	unavailability
Q	:	unreliability
P	:	probability
s	:	scheduled
f	:	force
R	:	reliability

CHAPTER-1

INTRODUCTION

General

Electricity has been the driving force for economies of the world and has become a daily provides day-to-day necessity for the population in the world. as its availability is linked with quality of life. Due to the nature of electricity systems, the variable demand at every moment needs to be met by consistent electricity supply to make sure the continuous availability of the resources. Not meeting the demand in any case lead to a huge loss of income to the generators as well as to the consumers. The reliability of the generation, transmission and distribution of electricity in this sense is crucial for the continuous supply of electricity to meet the demand.

Hydropower stations are one of prime contributors to the generating systems in most of the counties. Therefore, its role is crucial in supplying continuous and reliable power on customer's demand. The trend of reliability analysis in defense sector, aerospace sectors, nuclear reactors, communication sectors have been increasing day by day all over the world, but hydro professionals of developing countries have not yet given significant attention towards reliability analysis of hydropower stations. As a result, unplanned interruptions in power supply and load shedding are most common in most of the developing countries. Reliability analysis therefore is essential in decision making for operation, control, maintenance and operating of existing hydroelectric stations.

Reliability is a characteristic of an item, expressed by the probability that the item will perform its required function under given conditions for a stated time interval. It is generally designated by *R*. From a qualitative point of view, reliability can be defined as the ability of the item to remain functional[1]. Quantitatively, reliability specifies the probability has no operational interruptions will occur during a stated time interval. This does not mean that redundant parts may not fail; such parts can fail and be repaired (without operational interruption at item

Introduction

(system) level). The concept of reliability thus applies to non repairable as well as to repairable items[1].

Reliability of a hydropower stations basically depend on the availability of individual generating units and water availability for its operation. Availability of individual turbine-generating units depends on the availability of different components and subcomponents. Time taken to repair after occurrence of sudden and catastrophic failures of different equipment's is a major parameter in reliability evaluation. Based on the number of occurrence of failures and total repair time, mean time to repair (MTTR), mean time between failures (MTBF), mean time to failure (MTTF), failure rate, probability of occurrence can be estimated[2][3].

The Markov model can be used for a wide range of reliability problem including systems that are either non-repairable or repairable and are either series –connected, parallel redundant or standby redundant. A Markov process is a stochastic process in which at any given time the subsequent course of the process is affected only by the state at the given time and does not depend on the character of the process at any preceding time. Therefore, the accuracy of predicting the future of random process is not dependent on any knowledge or the extent of data on the past behavior of the process[4][5].

Pathri hydro power station (PHPS) is a canal based run –of river type hydro station with an installed capacity of 20.4MW.It consists of 3 identical unit of 6.8MW capacity per each. PHPS has been constructed on upper Ganga canal at 13 km downstream of holy city of Haridwar India.

Chilla hydro power station (CHPS) has an installed capacity of 144MW. It consists of 4 identical units of 36 MW capacities each. CHPS is a runoff river scheme constructed under Garhwal Rishikesh Chilla hydel scheme in the river Ganga. It comprises a diversion barrage across the river Ganga at Pashulok 5 km downstream of Rishikesh town. Each unit of CHPS comprises vertical shaft Kaplan turbine with rated head of 32.5 meter.

The objective of Evaluating Availability and Reliability will

A. Play an important role of knowing performance, ability, and weakness of each unit in order to plan preventive maintenance schedule.

Introduction

B. Help planning and deciding periodical maintenance, replacing or repairing when failure occurs.
C. Help in inventory control of spare parts.

Reliability studies are conducted for two purposes. Long-term evaluations are performed to assist in system planning and short-term evaluations to assist in day to day operating decisions. In short, these reliability indices (for long-term evaluations) are used by system planners and the authorities to decide on and advice for new investments in building new generation capacities[4].

Organization of the book:

Chapter 1 Emphasize the importance of reliability evaluation in general, followed by associated difficulties so as to define the objectives and the purpose of study.

Chapter 2 Deals with the fundamental of reliability, Basic concept and definition reliability, availability, unavailability, maintainability and many other reliability parameter like MTTR, MTBF, MTTF, failure rate, repair rate, force outage ,scheduled outage, failure density function.

Chapter 3 Concept of Marov modeling of a system. It highlight on different methods of reliability Markov model estimation.

Chapter 4 Present case study of PHPS and CHPS for its reliability and availability evaluation of methodology and its mathematical background.

Chapter 5 Present case study of PHPS for reliability and availability evaluation by using markov model, state space diagram of station, plant modeling ,unit modelling determination of failure rate, repair rate, state probabilities, reliability and availability.

Chapter 6 Present case study of CHPS for reliability and availability evaluation by using markov model, state space diagram of station, plant modeling ,unit modelling determination of failure rate, repair rate, state probabilities, reliability and availability.

Chapter 7 Results, discussion and valuable conclusion made during this study.

CHAPTER-2
FUNDAMENTAL OF RELIABILITY

2.1 Reliability

Reliability is a characteristic of an item, expressed by the probability that the item will perform its required function under given conditions for a stated time interval. It is generally designated by R. From a qualitative point of view, reliability can be defined as the ability of the item to remain functional. Quantitatively, reliability specifies the probability hat no operational interruptions will occur during a stated time interval. This does not mean that redundant parts may not fail; such parts can fail and be repaired (without operational interruption at system level). The concept of reliability thus applies to non repairable as well as to repairable items. To make sense, a numerical Statement of reliability must be accompanied by the definition of the required function, the operating conditions, and the mission duration. In general, it is also important to know whether or not the item can be considered new when the mission starts [4]. An item is a functional or structural unit of arbitrary complexity (e.g. component, assembly, equipment, subsystem, system) that can be considered as an entity for investigations. It may consist of hardware, software, or both and may also include human resources. Often, ideal human aspects and logistic Support are assumed, even if (for simplicity) the term System is used instead of technical system[6].

The Probability that a system or component will perform its intended function for a specified period of time, under required conditions. It can be also defined, as the probability that system, subsystem, or component will give specified performance for the duration for the duration of a mission when used in the manner and for the purpose intended, given that the system, subsystem, component is functioning properly at the start of the mission[7].

The term reliability is broad in meaning. In general, reliability designates the ability of a system to perform its assigned function, where past experience helps to form advance estimates of future Performance. a useful definition that illustrates the different dimensions of the concept is the following:

"Reliability is the probability of a device or system performing its function adequately, for the period of time intended, under the operating conditions intended "[7].

Reliability can be measured through the mathematical concept of probability by identifying the Probability of successful performance with the degree of reliability. Generally, a device or System is said to perform satisfactorily if it does not fail during the time of service. On the other hand, a broad range of devices are expected to undergo failures, be repaired and then returned to Service during their entire useful life[8]. In this case a more appropriate measure of reliability is the availability of the device, which is defined as follows:

"The availability of a repairable device is the proportion of time, during the intended time of service, that the device is in, or ready for service"[7].

The indices used in reliability evaluation are probabilistic and consequently, they do not provide exact predictions. They state averages of past events and chances of future ones by means of most frequent values and long-run averages. This information should be complemented with other economic and policy considerations for decision making in planning, design and operation[10].

2.2 Power System Reliability

System reliability is commonly interpreted as the probability of that system staying in the operating state, performing its intended purpose adequately for a period of time without failures under required conditions. The concept of power system reliability is extremely broad and covers all aspects of ability of system to satisfy the customer requirements[11]. The term applied to power system can be subdivided into two domains of adequacy and security assessment, as shown in figure 1.1.

Adequacy considers the system in static conditions which does not include system disturbances. It relates to the existence of sufficient facilities within the system to satisfy the consumer load demand. This includes necessary facilities to generate sufficient

Fundamental of reliability

electrical energy and associated transmission and distribution required to transfer energy to actual customer load points. It is the property of having enough capacity to remain secure almost all the time. In terms of generation, an adequate generation system is a matter of installed capacity and ability to meet the annual peak demand with this capacity under normal operating conditions, taking into account Scheduled and reasonably forced outages of generators. Adequacy, therefore, involves steady state post outage analysis of power systems. System Security, on other hand, relates to the ability of the system to withstand sudden perturbations arising within it. This includes the conditions associated with both local and widespread disturbances and loss of major generation and transmission facilities [12].

In terms of generation, generation system security is the capability of the generators in enduring unexpected contingencies involving frequency and voltage any time during system operation. Security is a dynamic measure of response to the unforeseen events. Security, therefore, involves the analysis of both static and dynamic conditions [13][14]. Together, adequacy and security provides the overall reliability description of the Power system, which can be broadly described as the ability to supply the quantity and quality of electricity desired by the customer when it is needed.

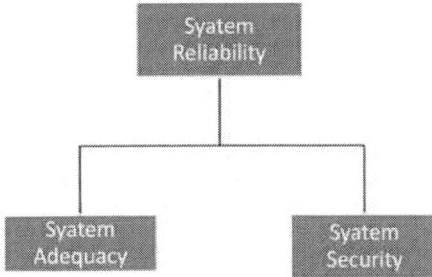

Fig.2.1: Components of system reliability

Electric power systems are extremely complex. This is due to many factors, some of which are sheer physical size, widely dispersed geography, national and international interconnections, flows that do not readily follow the transportation routes wished by operators but naturally follow physical laws, the fact that electrical energy cannot be stored efficiently or effectively in large quantities, unpredicted system behavior at one point of the system can have a major impact at large distances from the source of trouble, and many other reasons[15]. The historical development of and current scenarios within power companies is, however, relevant to an appreciation of why and how to evaluate the reliability of complex electric power systems.

Power systems have evolved over decades. Their primary emphasis has been on providing a reliable and economic supply of electrical energy to their customers. Spare or redundant capacities in generation and network facilities have been inbuilt in order to ensure adequate and acceptable continuity of supply in the event of failures and forced outages of plant, and the removal of facilities for regular scheduled maintenance[16]. The degree of redundancy has had to be commensurate with the requirement that the supply should be as economic as possible.

The probability of consumers being disconnected for any reason can be reduced by increased investment during the planning phase, operating phase, or both. Overinvestment can lead to excessive operating costs which must be reflected in the tariff structure. Consequently, the economic constraint can be violated although the system may be very reliable. On the other hand, underinvestment leads to the opposite situation. It is evident therefore that the economic and reliability constraints can be competitive, and this can lead to difficult managerial decisions

at both the planning and operating phases[17][18].

These problems have always been widely recognized and understood, and design, planning, and operating criteria and techniques have been developed over many decades in an attempt to resolve and satisfy the difference between the economic and reliability constraints. The criteria and techniques first used in practical applications, however, were all deterministically based. Typical criteria are[13][15]:

(a) Planning generating capacity—installed capacity equals the expected maximum demand plus a fixed percentage of the expected maximum demand;

(b) Operating capacity—spinning capacity equals expected load demand plus a reserve equal to one or more largest units;

(c) Planning network capacity—construct a minimum number of circuits to a load group and the minimum number being dependent on the maximum demand of the group.

Although these and other similar criteria have been developed in order to account for randomly occurring failures, they are inherently deterministic. Their essential weakness is that they do not and cannot account for the probabilistic or stochastic nature of system behavior, of customer demands or of component failures.

2.3 Probabilistic reliability criteria

System behavior is stochastic in nature, and therefore it is logical to consider that the assessment of such systems should be based on techniques that respond to this behavior (i.e., probabilistic techniques). It remains a fact, however, that most of the present planning, design, and operational criteria are based on deterministic techniques. These have been used by utilities for decades, and it can be, and is, argued that they have served the industry extremely well in the past. However, the justification for using a probabilistic approach is that it instills more objective assessments into the decision making process. In order to reflect on this concept it is useful to step back into history and recollect two quotes[13]:

A fundamental problem in system planning is the correct determination of reserve capacity. Too low a value means excessive interruption; while too high a value results in excessive costs. The greater the uncertainty regarding the actual reliability of any installation the greater the investment wasted.

The complexity of the problem, in general makes it difficult to find an answer to it by rules of thumb. The same complexity, on one side, and good engineering and sound economics, on the other, justify the use of methods of analysis permitting the systematic evaluations of all important factors involved. There are no exact methods available which permit the solution of reserve problems with the same exactness with which, say. Circuit problems are solved by applying Ohm's law.

The capacity benefits that result from the interconnection of two or more electric generating systems can best and most logically be evaluated by means of probability methods, and such benefits are most equitably allocated among the systems participating in the interconnection by means of "the mutual benefits method of allocation," since it is based on the benefits mutually contributed by the several systems[16].

2.4 Statistical and probabilistic measures

It is important to the point on what can be done regarding reliability assessment and why it is necessary. Failures of components, plant, and systems occur randomly; the frequency, duration, and impact of failures vary from one year to the next. There is nothing novel or unexpected about this. Generally all utilities record details of the events as they occur and produce a set of performance measures. These can be limited or extensive in number and concept and include such items as[11]:
• System availability;
• Estimated unsupplied energy;
• Number of incidents;
• Number of hours of interruption;
• Excursions beyond set voltage limits;
• Excursions beyond set frequency limits.

These performance measures are valuable because they:
    (a) Identify weak areas needing reinforcement or modifications;
    (b) Establish chronological trends in reliability performance;
    (c) Establish existing indices which serve as a guide for acceptable values in future reliability assessments;
    (d) Enable previous predictions to be compared with actual operating experience;
    (e) Monitor the response to system design changes.

The important point to note is that these measures are statistical indices. They are not deterministic values but at best are average or expected values of a probability distribution.

The same basic principles apply if the future behavior of the system is being assessed. The assumption can be made that failures which occur randomly in the past will also occur randomly in the future and therefore the system behaves probabilistically, or more precisely, stochastically. Predicted measures that can be compared with past performance measures or indices can also be extremely beneficial in comparing the past history with the predicted future[13]. These measures can only be predicted using probabilistic techniques and attempts to do so using deterministic approaches are delusory.

In order to apply deterministic techniques and criteria, the system must be artificially constrained into a fixed set of values which have no uncertainty or variability. Recognition of this restriction results in an extensive study of specified scenarios or "credible" events. The essential weakness is that likelihood is neglected and true risk cannot be assessed[6].

At this point, it is worth reviewing the difference between a hazard and risk and the way that, these are assessed using deterministic and probabilistic approaches. The two concepts, hazard and risk, are often confused; the perception of a risk is often weighed by emotion which can leave industry in an invidious position. A hazard is an event which, if it occurs, leads to a dangerous state or a system failure. In other words, it is an undesirable event, the severity of which can be ranked relative to other hazards. Deterministic analyses can only consider the outcome and ranking of hazards. However, a hazard, even if extremely undesirable, is of no consequence if it cannot occur or is so unlikely that it can be ignored. Risk, on the other hand, takes into account not only the hazardous events and their severity, but also their likelihood. The combination of severity and likelihood creates plant and system parameters that truly represent risk. This can only be done using probabilistic techniques[5].

2.5 Absolute and relative measures

It is possible to calculate reliability indices for a particular set of system data and conditions. These indices can be viewed as either absolute or as relative measures of system reliability. Absolute indices are the values that a system is expected to exhibit. They can be monitored in terms of past performance because full knowledge of them is known. However, they are extremely difficult, if not impossible, to predict for the future with a very high degree of confidence. The reason for this is that future performance contains considerable uncertainties particularly associated with numerical data and predicted system requirements. The models used are also not entirely accurate representations of the plant or system behavior but are approximations. This poses considerable problems in some areas of application in which absolute values are very desirable. Care is therefore vital in these applications, particularly in situations in which system dependencies exist, such as common cause (mode) failures which tend to enhance system failures.

Relative reliability indices, on the other hand, are easier to interpret and considerable confidence can generally be placed in them. In these cases, system behavior is evaluated before and after the consideration of a design or operating change. The benefit of the change is obtained by evaluating the relative improvement. Indices are therefore compared with each other and not against specified targets. This tends to ensure that uncertainties in data and system requirements are embedded in all the indices and therefore reasonable confidence can be placed in the relative differences[13]. In practice, a significant number of design or operating strategies or scenarios are compared, and a ranking of the benefits due to each is made. This helps in deciding the relative merits of each alternative, one of which is always to make no changes.

The most important aspect to remember when evaluating these measures is that it is necessary to have a complete understanding of the engineering implications of the system. No amount of probability theory can circumvent this important engineering function. It is evident therefore that probability theory is only a tool that enables an engineer to transform knowledge of the system into a prediction of its likely future behavior. Only after this understanding has been achieved can a model be derived and the most appropriate evaluation technique chosen. Both the model and the technique must reflect and respond to the way the system operates and fails. Therefore the basic steps involved are[15]:

• understand the ways in which components and system operate;
• identify the ways in which failures can occur;
• deduce the consequences of the failures;
• derive models to represent these characteristics;
• only then select the evaluation technique.

2.6 Methods of assessment

Power system reliability indices can be calculated using a variety of methods. The two main approaches are analytical and simulation. The vast majority of techniques have been analytically based and simulation techniques have taken a minor role in specialized applications. The main reason for this is because simulation generally requires large amounts of computing time, and analytical models and techniques have been sufficient to provide planners and designers with the results needed to make objective decisions. This is now changing, and increasing interest is being shown in modeling the system behavior more comprehensively and in evaluating a more informative set of system reliability indices[4].

Analytical techniques represent the system by a mathematical model and evaluate the reliability indices from this model using direct numerical solutions. They generally provide expectation indices in a relatively short computing time. Unfortunately, assumptions are frequently required in order to simplify the problem and produce an analytical model of the system. This is particularly the case when complex systems and complex operating procedures have to be modeled. The resulting analysis can therefore lose some or much of its significance. The use of simulation techniques is very important in the reliability evaluation of such situations.

Simulation methods estimate the reliability indices by simulating the actual process and random behavior of the system. The method therefore treats the problem as a series of real experiments. The techniques can theoretically take into account virtually all aspects and contingencies inherent in the planning, design, and operation of a power system. These include random events such as outages and repairs of elements represented by general probability distributions, dependent events and component behavior, queuing of failed components, load variations, variation of energy input such as that occurring in hydro generation, as well as all different types of operating policies[13].

If the operating life of the system is simulated over a long period of time, it is possible to study the behavior of the system and obtain a clear picture of the type of deficiencies that the system may suffer. This recorded information permits the expected values of reliability indices together with their frequency distributions to be evaluated. This

comprehensive information gives a very detailed description, and hence understanding, of the system reliability. The simulation process can follow one of two approaches[7]:
- (a) Random—this examines basic intervals of time in the simulated period after Choosing these intervals in a random manner.
- (a) Sequential—this examines each basic interval of time of the simulated period in chronological order.

The basic interval of time is selected according to the type of system understudy, as well as the length of the period to be simulated in order to ensure a certain level of confidence in the estimated indices.

The choice of a particular simulation approach depends on whether the history of the system plays a role in its behavior. The random approach can be used if the history has no effect, but the sequential approach is required if the past history affects the present conditions. This is the case in a power system containing hydro plant in which the past use of energy resources (e.g., water) affects the ability to generate energy in subsequent time intervals. It should be noted that irrespective of which approach is used, the predicted indices are only as good as the model derived for the system, the appropriateness of the technique, and the quality of the data used in the models and techniques.

2.7 Basic definition related to reliability

2.7.1 Failure

A failure occurs when the item stops performing its required function. As simple as this definition is, it can become difficult to apply it to complex items. The failure free time is generally random variable. It is often reasonably long, but it can be very short, for instance because of a failure caused by a transient event at turn-on. A general assumption in investigating failure-free times is that at $t = 0$ the item is free of defects and systematic failures. Besides their frequency, failures should be classified according to the mode, cause, effect, and mechanism[9]:

1. *Mode:* The mode of a failure is the Symptom (local effect) by which a failure is observed; e.g., Opens, shorts, or drift for electronic components; brittle rupture, creep, cracking, seizure, fatigue for mechanical components.
2. *Cause:* The cause of a failure can be intrinsic, due to weaknesses in the item and/or wear out, or extrinsic, due to errors, misuse or mishandling during the design, production, or use. Extrinsic causes often lead to systematic failures, which are deterministic and should be considered like defects (dynamic defects in software quality). Defects are present at $t = 0$, even if often they cannot be discovered at $t = 0$. Failures appear always in time, even if the time to failure is short as it can be with systematic or early failures.
3. *Effect:* The effect (consequence) of a failure can be different if considered on the item itself or at higher level. A usual classification is: non relevant, partial, complete, and critical failure. Since a failure can also cause further failures, distinction between primary and secondary failure is important.
4. *Mechanism:* Failure mechanism is the physical, chemical, or other process resulting in a failure.

Failures can also be classified as sudden and gradual. In this case, sudden and complete failures are termed cataleptic failures, gradual and partial failures are termed degradation failures. As failure is not the only cause for an item being down, the general term used to define the down state of an item (not caused by a preventive maintenance, other planned actions, or lack of external resources) is fault. Fault is thus a state of an item and can be due to a defect or a failure[9].

In order to apply deterministic techniques and criteria, the system must be artificially constrained into a fixed set of values which have no uncertainty or variability. Recognition of this restriction results in an extensive study of specified scenarios or "credible" events. The essential weakness is that likelihood is neglected and true risk cannot be assessed. At this point, it is worth reviewing the difference between a hazard and risk and the way that, these are assessed using deterministic and probabilistic approaches.

2.7.2 Failure Density Function

Failure can be defined as the termination of the ability of an item to perform its required function. Components failures are generally classified as either catastrophic failures or degradation failures. Degradation failures are sometimes called parametric drift failures. Catastrophic failures are both sudden and complete. A sudden failure is one, which cannot be anticipated, and a complete failure is one which results in complete cessation of the function required. Degradation failures are both gradual and partial and result in deviations from acceptable limits without complete cessation of the function required. A complete failure is an extreme case of partial failure. The failure of a component is random in nature, various probability distributions are of great significance in order to derive hazard functions. Some of the commonly used probability distributions in reliability analysis are Exponential, Binomial, Poisson, Normal, Gamma, Weibull, etc. However in most of the reliability analysis studies, exponential distribution is the most widely used. The probability density function or failure density function, f(t) in exponential distributions defined as[3]

$$f(t)=$$

Where $t \geq 0$ and λ is the hazard rate or failure rate

2.7.3 Forced outage:

An outage result from emergency conditions directly associated with component or unit requiring that unit be taken out of service immediately, either automatically or as soon as switching operations can be preformed[14].

2.7.4 Scheduled outage:

An outage that results when unit is deliberately taken out of service at selected time, usually for purpose of construction, preventive maintenance, repair or reserve[14].

2.7.5 Maintenance, Maintainability

Maintenance defines the set of activities performed on an item to retain it in or to restore it to a specified state. Maintenance is thus subdivided into preventive maintenance, carried out at predetermined intervals to reduce wear out failures, and corrective maintenance, carried out after failure recognition and intended to put the item into a state in which it can again perform the required function. Aim of a preventive maintenance is also to detect and repair hidden failures, i.e. failures in redundant elements not identified at their occurrence. Corrective maintenance is also known as repair, and can include any or all of the following steps: recognition, isolation (localization & diagnosis), elimination (disassembly, replace, reassembly),checkout. Repair is used hereafter as a synonym for restoration. To simplify calculations, it is generally assumed that the element in the reliability block diagram for which a maintenance action has been performed is as-good-as-new after maintenance. This assumption is valid for the whole equipment or system in the case of constant failure rate for all elements which have not been repaired or replaced [2].

Maintainability is a characteristic of an item, expressed by the probability that a preventive maintenance or a repair of the item will be performed within a stated time intently for given procedures und resources (skill level of personnel, spare Parts, test facilities, etc.). From a qualitative point of view, maintainability can be defined as the ability of an item to be retained in or restored to a specified state.

Maintainability is defined as the probability that a device or system that has failed will be restored to operational effectiveness within a given time. The time to repair includes several activities, usually divided into three groups.

• Preparation time: finding the person for the job, travel, obtaining tools and test equipments, etc.

• Active maintenance time: actually doing the job

• Delay time (logistic time): waiting for spares, etc., once the job has been started.

The time taken to repair failures and to carry out routine preventive maintenance removes the system from the available state. Thus, there is a close relationship between reliability and maintainability, one affecting the other and both affecting availability and costs.

2.7.6 Availability and Unavailability

Availability is defined as the probability that an item will be available when required the proportion of the total time the item is available for use or "the proportion of time, in the long run that is in or ready for, service"[14].

Availability is a broad term, expressing the ratio of delivered to expected service. It is often designated by A and used for the stationary and steady-state value of the point and average availability. Point availability is a characteristic of an item expressed by the probability that the item will perform its required function under given conditions at a stated instant of time t. From a qualitative point of view, point availability can be defined as the ability of the item to perform its required function under given conditions at a stated instant of time (dependability).Availability evaluations are often difficult, as logistic support and human factors should be considered in addition to reliability and maintainability. Ideal human and logistic support conditions are thus often assumed, yielding to the intrinsic (inherent) availability. Hereafter, availability is used as a synonym for intrinsic availability. Further assumptions for calculations are continuous operation and complete renewal for the repaired element in the reliability block diagram. Therefore, the availability of a repairable item is a function of its failure rate (λ) and repair rate (μ). The proportion of the total time that the item is available is the steady state availability . Therefore, the steady state availability is given by[2]

Fundamental of reliability

Then A ——

2.7.7 Mean Time to Failure (MTTF), Mean Time between Failures (MTBF) and Mean Time to Repair (MTTR)

MTTF and MTBF are often two confusing terms in reliability engineering. MTTF is the mean Operating time between Successive failures where as MTBF is simply the mean time between failures.

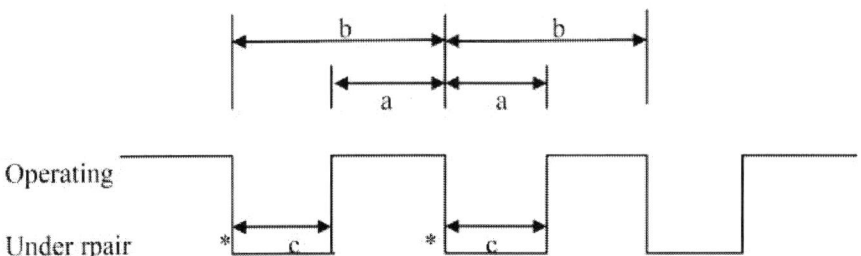

Fig2.2: MTBF V$_s$ MTTF

MTBF = Mean of time values 'b'

MTTF = Mean of time values 'a'

MTTR = Mean of time values 'c'

* = Time at which failure occurs

MTTR is the mean of the repair time of a repairable component or a system

Fundamental of reliability

$$MTTF + MTTR = MTBF$$

MTTF and MTBF are more important in the case of repairable components or system But sometimes in the case of repairable components or systems too, when repair time is very small compared to the operating time, the numerical value of MTTF and MTBF are very close. But if the repair time is long, two terms MTTF and MTBF are Significantly different. Thus in the case of non-repairable components or systems, these terms MTTF and MTBF become similar[3].

2.7.8 Failure Rate and Repair Rate

Failure rate is defined as the reciprocal of MTTF and repair is defined as the reciprocal of MTTR.

If $\quad m = MTTF$

$\quad\quad r = MTTR$

Then, failure rate (λ) and repair rate (μ) is expressed as

Failure rate, $\lambda \quad\rule{1cm}{0.15mm}\quad\rule{0.5cm}{0.15mm}$

Repair rate, $\mu \quad\rule{1cm}{0.15mm} =\rule{0.3cm}{0.15mm}$

Here, the failure rate and repair rate have been assumed as constant and thus not expressed with suffix time t.

The probability of residing of the system or component in operable stale, that is, reliability R and the Probability of not residing the component or the system that is, unreliability Q are expressed as

$$R \quad \frac{\mu}{\mu+} \quad \rule{1cm}{0.15mm}$$

29

$$Q \quad — \quad —$$

From above equation

$$R + Q = 1$$

2.7.9 Cycle Time and Cycle Frequency

Since system operating (i.e. up time) and the system in repair (i.e. down time) forms a cycle, the sum of MTTF and MTTR is called as the system cycle time T

$$T = MTBF = MITF + MTTR$$

And, the reciprocal of the system cycle time is known as system cycle frequency or simply cycle frequency f

$$f \quad -$$

In other way, cycle frequency can be understood as the frequency of encountering a system state. In the case of two State system, the frequency of encountering the operating state is same as that of encountering the failed state

2.8 Reliability Block Diagram (RBD)

A Reliability Block Diagram (RBD) performs the system reliability and availability analyses on large and complex systems using block diagrams to show network relationships. The structure of the reliability block diagram defines the logical interaction of failures within a system that are required to sustain system operation. The rational course of a RBD stems from an input node located at the left side of the diagram. The input node flows to arrangements of series or parallel blocks that conclude to the output node at the right side of the diagram. A diagram should only contain one input and one output node. The RBD system is connected by a parallel or series configuration.

In setting up the reliability block diagram, care must be taken regarding the fact that only two states (good or failed) and one failure mode (e.g., Opens or shorts) can be considered for each element. Particular attention must also be paid to the correct identification of the parts which appear in series with a redundancy[1].

2.8.1 System Reliability Models

The different types of reliability models generally encountered during reliability evaluation are mentioned in the following sections.

Series System:

The components in a set are said to be in series form from reliability point of view if they must all work for a system success or only one need to fail for system failure

Fig 2.3: Two series component

Let two components A and B are connected in series from reliability point of view that both components must work to ensure the success of the system.

Let R_a = Probability of successful operation of component A

R_b = Probability of successful operation of component B

And Q_a = Probability of failure of component A

Q_b = Probability of failure of component B

Since success and failure are mutually exclusive events and complementary events,

$$R_a + Q_a = 1$$

$$R_b + Q_b = 1$$

From theory of probability, we can derive as

Probability of success of series system $Rs = R_a \cdot R_b$

Fundamental of reliability

if n number of components are in series then

System success probability

$$Rs =$$

Here Ri like symbol represent the multiplication

Thus, failure probability of the system, $Qs = 1 - Rs = 1 -$

Parallel System:

The components are said to be in parallel from reliability point of view if only One needs to be working for system success or all must fail for system failure

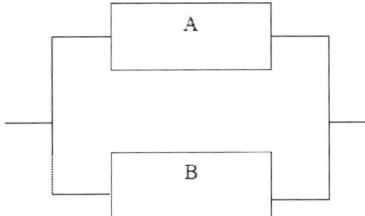

Fig 2.4: Two component parallel system

In the system, composed of two independent components A and B with reliabilities R_a and Q_a the system operates satisfactory if either one or both components function. Therefore, from the theory of probability we can derive

The probability of failure of the system $Qp = Q_a.Q_b$

Where $Q_a = 1-R_a$ and

When n number of components in parallel

The equivalent reliability of components A and B in series is $Rab = R_a.R_b$

The equivalent reliability of components A and B in series is $Rcd = R_c.R_d$

Now, the failure probability of the resulting parallel system can be found as

Fundamental of reliability

$$Qs = Qab.Qcd = (1-Rab).(1-Rcd)$$

Thus, the reliability of the system, $Rs = 1- \{(1-Ra.Rb).(1-(Rc.Rd)\}$

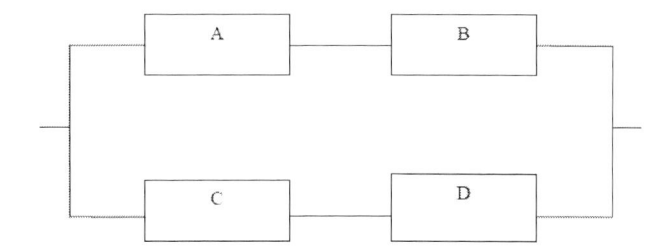

Fig 2.5:Series – parallel system

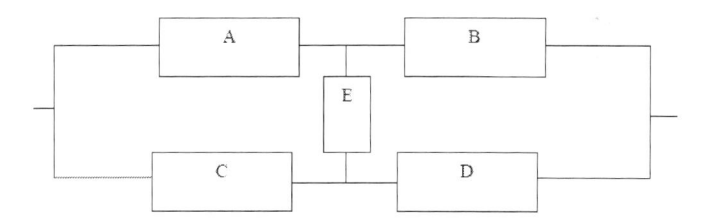

2.8.2 Non-series- parallel system :

Consider a bridge type of network, to which, everybody might have been familiar to. The fig below shows figure 1.6 a bridge type of network. In this network indicates none o the components is connected in simple series/parallel arrangement. This type of system or network is generally called as nonseries/parallel networks or complex networks. There are a number of techniques available for solving such type of network and for more complex networks as well.

Fig2.6:Non-Series- Parallel system

CHAPTER -3
MARKOV MODELLING AND ESTIMATION

3.1 Introduction:

Reliability problems are normally concerned with systems that are discrete in space, i.e., they can exist in one of a number of discrete and identifiable states, and continuous in time, i.e., they exist continuously until another transition occurs. The Markov model can be used for a wide range of reliability problem including systems that are either non-repairable or repairable and are either series –connected, parallel redundant or standby redundant [9]. A Markov process is a stochastic process in which at any given time the subsequent course of the process is affected only by the state at the given time and does not depend on the character of the process at any preceding time. Therefore, the accuracy of predicting the future of our random process is not dependent on any knowledge or the extent of data on the past behavior of the process [10] .A large number of systems of engineering interest can be described by a Markov process, by a correct choice of state variable. Like stochastic process, in general, Markov process can be divided into discrete parameter and continuous parameter process. If time is a discrete event, we normally call the Markov Process a Markov chain. Therefore, discrete or continuous parameter Markov chains, in which time is discrete and discrete or continuous parameter Markov processes in which time is continuous.

3.2 Continuous Markov Process

3.2.1 Transition rate concepts

Consider the case of single repairable component for which the failure rate and repair rate are constant, i.e. they are characterized by the exponential distribution. The state transition diagram for this component is shown in fig.4.1

The failure density function for a component with a constant hazard rate of λ was given in

Markov modelling and estimation

.... (3.1)

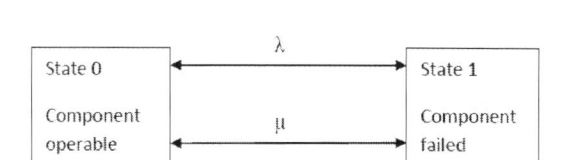

```
          λ
┌─────────────┐ ←──────→ ┌─────────────┐
│   State 0   │          │   State 1   │
│  Component  │    μ     │  Component  │
│  operable   │ ←──────→ │   failed    │
└─────────────┘          └─────────────┘
```

Fig.3.1 Single component repairable system-State Space Diagram

Where,

λ = failure rate = ─────────────────────────

μ = repair rate = ─────────────────────────

The parameter λ and μ are referred to as state transition rates since they represent the rate at which the system transits from one state of the system to another [5].

3.2.2 Evaluating time dependent probabilities:

The relevant state space diagram for the simple single component is shown in fig.. In the case of continuous Markov processes they are usually represented by a transition rate as shown in above figure, by the transition λ and μ from the operating and failed states respectively. Consider now an incremental interval of time dt which is made sufficiently small so that the probability of two or more events occurring during this increment of time is negligible.

$P_0(t) = \text{───} + \text{─────}$ .... (3.2)

$P_1(t) = \text{───} - \text{─────}$ .... (3.3)

The probability $P_0(t)$ and $P_1(t)$ are the probabilities of being found in the operating state and failed state respectively as a function of time given that the system started at time t = 0 in the operating state[1].

3.2.3 Evaluating limiting state probabilities

The limiting state or steady state probabilities will be non zero for a continuous Markov process provided the system argotic, as in the case of a discrete Markov chain. In case of simple repairable component represented by the state space diagram shown in above figure. The limiting probability can be evaluated from equation (ii) and (iii) by letting t → ∞.If these values of limiting state probabilities are defined as P_0 and P_1 for the operating state and the failed state ,respectively, then from equation (ii) and (iii) as t → ∞

$$P_0 = P_0 (\infty) = \text{———} \quad \text{..........................(3.4)}$$

$$P_1 = P_1 (\infty) = \text{———} \quad \text{..... ....(3.5)}$$

These limiting state probability expressions are applicable irrespective of whether the system starts in the operating state or in the operating state or in failed state. The value of P_0 and P_1 are generally referred to as the steady state or limiting availability A and unavailability U of the system respectively. The time dependent availability A(t) of the system is given by equation (ii) and (iii) i.e.,

$$A(t) = P_0(t) = \text{———} + \text{————} \quad \text{..................(3.6)}$$

This is probability being found in the operating state at some time t in the future given that the system started in the operating state at time t = 0. This is quite different from the reliability R(t) as given by

$$R(t) = e^{-\lambda t} \quad \text{................................(3.7)}$$

This is the probability of staying in the operating as a function of time given that the system started in the operating state at time t = 0.The limiting state probabilities could have been evaluated directly and simply from the differential equation.

3.3 State Space Diagrams

The solution of continuous or discrete Markov processes it is desirable first to construct the appropriate state space diagram and insert the relevant transition rates. All relevant

Markov modelling and estimation

State in which the system can reside should be including in such a diagram and all known ways in which transition between states can occur should be inserted. There are no basic restrictions on the number of states or the type and number of transition that can be inserted. The analyst must therefore first translate the operation of the system into a state space diagram recognizing both the states of the way these states communicate and the values of the transition rates. It is relatively easy to formulate state space diagrams for small system models [5].

3.3.1 Single repairable component

Fig.3.1 shows a state space diagram for a single repairable component which is assumed to exist in one of two states, the operating, or up state and the failed, or down, state. In some practical situations, a single component may be best represented by more than two states. For example, a pump in good working order may be able to deliver full output, under certain conditions may only be able to deliver partial output (known as a partial output or derated state) and when failed is unable to deliver any output. This therefore gives three states as shown in fig.3.2. In a given practical application, additional derated states may exist and it is necessary for the analyst to appreciate these additional states and represent them in the diagram.

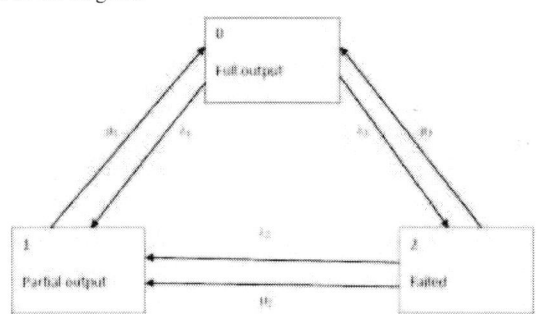

Fig.3.2 State Space diagram of component with partial output state

The diagram shown in figure 3.2 includes all possible transition rates. Some of these may not physically exist in practice, in which case they should be omitted from the diagram. One of the most likely transitions not to exist is the rate μ_2, since it is probable that, once failed, the repair process will return it to the full output state[1].

Markov modelling and estimation

3.3.2 Two repairable components

If the system consists of two repairable components, there are four possible states in which the system can exist. If λ_1, μ_1 and λ_2, μ_2 are the failure and repair rates of components 1 and 2 respectively, the state space diagram including the relevant transition rates is shown in Figure3.3.

One important feature to recognize in this particular example is that the state space diagrams shown in figure 3.3 is the same irrespective of whether the two components are in series or are parallel redundant. In the case of a series system, state 1 is the system up state and state 2, 3 and 4 are the system down states. In the case of a parallel redundant system, state 1, 2 and 3 are the system up states and state 4 is the system down state. Therefore if P_1, P_2, P_3 and P_4 are the probabilities of being in states 1-4 respectively[9]. Then

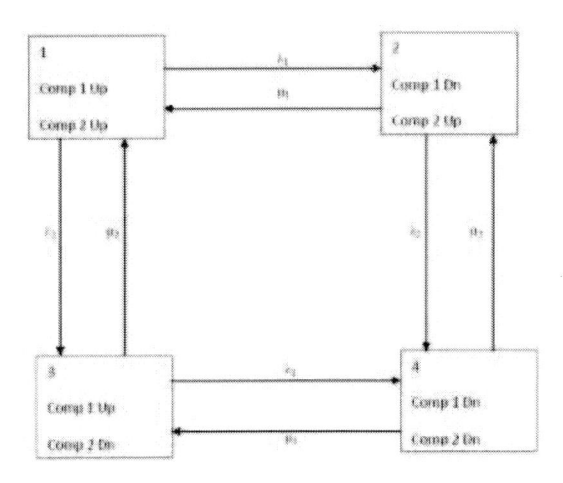

Fig.3.3 State space diagram for two component system

$$\left.\begin{array}{ll} \text{For series system} & P_{up} = A = P_1 \\ \\ & P_{down} = U = P_2 + P_3 + P_4 \\ \\ \text{For the parallel redundant system} & P_{up} = A = P_1 + P_2 + P_3 \end{array}\right\} \quad \ldots\ldots\ldots\ldots(3.8)$$

$$P_{down} = U = P_4$$

and these equations apply for both the time – dependent and limiting state probabilities. Certain transitions in the state space model may not be physically possible and should be deleted and other should perhaps be added. As an example, if both components of the system have failed, repair of component 2 may never be started until component 1 has been repaired in which case the transition μ_2 from state 4 to state 2 would not exist. It may also be possible for both components to fail simultaneously in which case a transition between state 1 and state 4 will exist. The physical transition from state 1 to state 4 would be via state 2 or 3.

In some practical situations, the state space diagram shown in figure 3.3 can be simplified and reduced. For example, it may be known that when one of the components fails in a series system, the other component is no longer operating and its failure rate in these circumstances becomes zero. In this case, state 4 does not exist leaving only states 1-3 and the transition rates between these 3 states [6].

If both of the components are identical, states 2 and 3 are also identical and may be combined to give a reduced 3- state system as shown in figure 3.4 The 2λ and 2μ terms in figure 3.4 indicate that two components are available for failure or repair respectively in the next increment of time and that one of the two can fail or be repaired, but not both in that interval.

3.3.3 Large number of components

The number of states in the state space diagrams increases as the number of system components increase and as the number of state in which each system component can reside increases .If all states are represented.

The number of states in the diagram is 2^n for an n-component system in which each component is represented as a 2-state model and 3^n if each component has a derated states .the model can therefore become unmanageable for large systems. Two solutions are possible in these circumstances. The first involves state truncation [1][9].

Markov modelling and estimation

This approach utilizes engineering judgment based on experience to reduce the number of possible system states by neglecting those that have a very low probability of occurrence. The second solution involves approximate solution technique based on Markov modeling.

3.3.4 Standby redundant system

The important class of systems known as standby system can also be modeled and analyzed using state space diagram and Marov technique. The most importance aspect in the representation of these systems is to recognize and identify the state in which the system can reside and the transition that can occur between these states. In order to illustrate the construction of state space diagrams reconsider the two component standby system shown in figure 3.6 in which the switch was assumed to be perfect.

The state space diagram of this system is shown in figure 3.6. In this diagram the subscripts o, f and s it is assumed that component A is the normally operating component. The diagram shown in figure 3.6 is not meant to be a rigorous or exhaustive representation. It assumes that, whenever A is operable, it will replace B as the operating component. In circumstances when A and B are identical, it may be operating policy to leave B operating and place A on standby to give a state equivalent to state 1 with the status of the components reversed. If the switch is not perfect, each state of the diagram must include the status of not only A and B but also of the switch itself which result in a greater number of states[1].

3.3.5 Mission oriented system

State space diagrams can however be constructed for non-repairable or mission oriented system in exactly the same way and can be solved using Markov technique. The only difference between these systems and repairable system is that the repair transitions of non-repairable components do not exist. The state space diagram of figure 3.4 for a system having two identical components is modified to that shown in figure 3.7, if the components are non-repairable[1].

Markov modelling and estimation

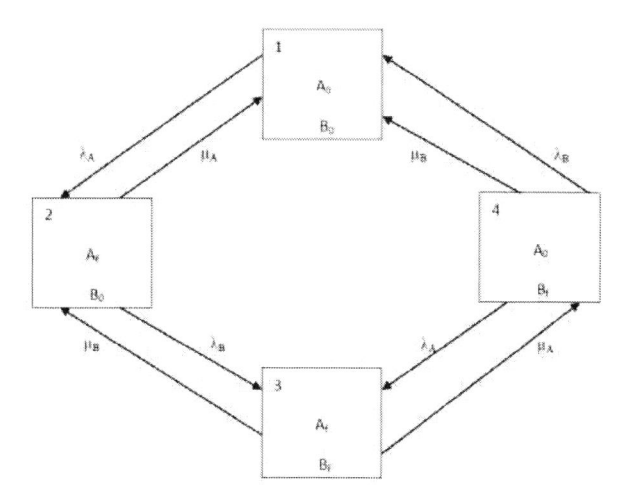

Fig.3.4 State space diagram for two component standby system

In this case, the system is no longer argotic since not all states can communicate and one state is an absorbing state.

The time dependent probabilities of non-repairable systems can be evaluated by using Markov process technique in the same way as for repairable systems .The limiting state probabilities However, do not have any significance since, in the limit ,the probability of residing in the absorbing state is unity and the probability of residing in all non-absorbing states is zero.

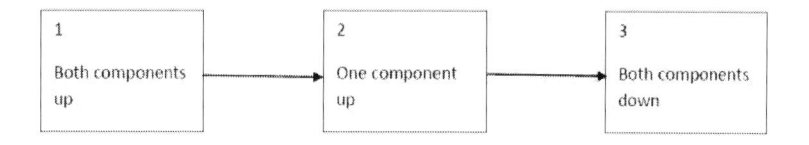

Fig.3.5: State Space diagram for two identical non-repairable components

3.4 Multi–state problems

3.4.1 Two component repairable system

The basic frequency and duration concept can be applied to any size of repairable system and have been applied in practice to power system generation problems containing several hundred individual units. The stochastic transitional probability matrix for the system is

$$P = \begin{array}{c} \\ 1 \\ 2 \\ 3 \\ 4 \end{array} \begin{pmatrix} \overset{1}{1-(\lambda_1+\mu_1)} & \overset{2}{\lambda_1} & \overset{3}{\lambda_2} & \overset{4}{-} \\ \mu_1 & 1-(\lambda_2+\mu_1) & - & \lambda_2 \\ \mu_2 & - & 1-(\lambda_1+\mu_2) & \lambda_1 \\ - & \mu_2 & \mu_1 & 1-(\lambda_2+\mu_2) \end{pmatrix}$$

The application of the technique to multi component systems can be considering a simple two component system, in which each component is considered to have an up state (operating) and a down state (failed) with failure and repair rates of λ_1, μ_1 and λ_2, μ_2 foe component 1 and 2 respectively. The state space diagram of this system shown in figure 3.3 is replaced in figure 3.8.

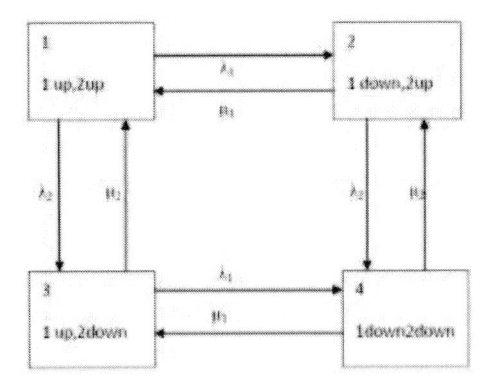

Fig.3.6 State space diagram for two component system

Markov modelling and estimation

3.4.2 State Probabilities

The first step in the frequency and duration method is to evaluate the individual limiting state probabilities. The state probabilities are

$$P_1 = \underline{\hspace{2cm}}$$

$$P_2 = \underline{\hspace{2cm}} \quad \ldots\ldots\ldots(3.9)$$

$$P_3 = \underline{\hspace{2cm}}$$

$$P_4 = \underline{\hspace{2cm}}$$

Since the identical states of the system are mutually exclusive, the probabilities given by above equation can be combined to give probability of residing in any set of cumulated states[13].

For example,

(a) For a series system, $P_{up} = P_1$

$$P_{down} = P_2 + P_3 + P_4$$

(b) For a parallel System, $P_{up} = P_1 + P_2 + P_3$

$$P_{down} = P_4$$

3.4.3 Frequency of encountering individual states

The second step in the evaluation of the frequency and duration indices of a system is to evaluate the frequency of encountering the individual states. This is obtained by above equation and individual state probabilities and the rate of departure or entry. The relevant state space diagram for the simple single component is shown in fig.. In the case of continuous Markov processes they are usually represented by a transition rate as shown in above figure, by the transition λ and μ from the operating and failed states respectively.

Table 3.1 the rate of departure or entry

State number	Component 1	Component 2	Rate of Departure	Rate of Entry
1	up	up	$\lambda_1 + \lambda_2$	$\mu_1 + \mu_2$
2	down	up	$\lambda_2 + \mu_1$	$\lambda_1 + \mu_2$
3	up	down	$\lambda_1 + \mu_2$	$\lambda_2 + \mu_1$
4	down	down	$\mu_1 + \mu_1$	$\lambda_1 + \lambda_2$

(a) Frequency of encountering state 1:

If f_1 = Frequency of encountering state 1

= $P_1 \times$ (rate of departure from state 1)

= ─────── ....(3.10)

(b) Frequency of encountering state 4:

If f_4 = Frequency of encountering state 4

= $P_4 \times$ (rate of departure from state 4)

= ─────── ....(3.11)

Similarly derivation may be made for f_2 and f_3, the frequency of encountering state 2 and 3 respectively. The complete list of state probabilities and frequencies of encounter are shown in Table 3.2.

In the case when both components are identical, that is $\lambda_1=\lambda_2=\lambda_3=\lambda$ and $\mu_1=\mu_2=\mu_3=\mu$

Markov modelling and estimation

$$f_1 = 2\lambda\mu^2 / (\lambda + \mu)^2$$
$$f_1 = 2\lambda^2\mu / (\lambda + \mu)^2 \qquad \biggr\} \quad \ldots\ldots(3.12)$$
$$f_2 = f_3 = \lambda\mu / (\lambda + \mu)$$

Table 3.2 State probabilities and frequencies

State number	Probability	Frequency of encounter
1	$\mu_1\mu_2 /D$	$\mu_1\mu_2 (\lambda_1 + \lambda_2)/D$
2	$\lambda_1\mu_2 /D$	$\lambda_1\mu_2 (\lambda_2 + \mu_1)/D$
3	$\mu_1\lambda_2 /D$	$\mu_1\lambda_2 (\lambda_2 + \mu_1)/D$
4	$\lambda_1\lambda_1 /D$	$\lambda_1\lambda_1 (\mu_1 + \mu_2)/D$

3.4.4 Cycle time between individual states:

The cycle time T defined as the reciprocal of the frequency of encounter f. a cycle for each individual state can therefore be deduced from the frequency of encounter this state. This value of cycle time represents the mean time between entering (and departing from) a given state to next entering (or next departing from) the same state. In the case of the two component system and considering the case of identical components

$$T_1 = \underline{\qquad}$$
$$T_2 = T_3 = \underline{\qquad} = - + - \qquad \ldots\ldots\ldots\ldots(3.13)$$
$$T_4 = \underline{\qquad}$$

Consider the two case of parallel redundant systems and series system.

45

Markov modelling and estimation

(a) Parallel redundant systems

The MTTF of a parallel system is defined as the mean time between encounters of the state in which both components are down.

$$MTTF = \text{———} \quad \quad \quad \quad \quad \quad \quad \quad (3.14)$$

Since the MTBF of a parallel redundant syatem is given by the sum of the MTTF anf MTTR, then

$$MTBF = \text{———} + \text{—}$$

$$MTBF = \text{————}$$

$$MTBF = \text{———} \quad \quad \quad \quad \quad \quad \quad \quad (3.15)$$

Which is identical equation, obtained from the cycle time T_4 using the frequency and duration method.

(b) Series systems

In the case of a series system the MTTF is $1/2\lambda$ and the MTTR is $(\lambda + 2\mu)/2\mu^2$. Therefore

$$MTBF = \text{———} + \text{—}$$

$$MTBF = \text{————}$$

$$MTBF = \text{———} \quad \quad \quad \quad \quad \quad \quad \quad (3.16)$$

Which is identical to the time cycle time T_1 given by above equation. Again these two results are expected to be identical since the up state of the series system is state 1 when both components are up, and the MTBF of the system is given by the mean time of encountering this time[11].

3.5 Discrete Markov Process:

The discrete-time process $\{X_k, k = 0, 1, 2, \ldots\}$ is called a Markov chain if for all i, j, k, . . . , m, the following is true

$$P[X_{k=j}, X_{k-1=i}, X_{k-2=n}, \ldots, X_{0=m}] = P[X_{k=j}|X_{k-1=i}] = p_{ijk} \quad \ldots(3.17)$$

The quantity pijk is called the state transition probability, which is the conditional probability that the process will be in state j at time k immediately after the next transition, given that it is in state i at time k − 1. A Markov chain that obeys the preceding rule is called a non homogeneous Markov chain[9]. We considered homogeneous Markov chains, which are Markov chains in which $p_{ijk} = p_{ij}$. This means that homogeneous Markov chains do not depend on the time unit, which implies that

$$P[X_{k=j}|X_{k-1=i}, X_{k-2=n}, \ldots, X_{0=m}] = P[X_{k=j}, X_{k-1=i}] = p_{ij} \quad \ldots\ldots(3.18)$$

The homogeneous state transition probability pij satisfies the following conditions

1. $0 \leq p_{ij} \leq 1$
2. $\quad = 1, i = 1, 2, \ldots, n,$

which follows from the fact that the states are mutually exclusive and collectively exhaustive From the definition we obtain the following Markov chain rule:

$$P[X_{k=j}, X_{k-1=i}, X_{k-2}, \ldots, X_0] = P[X_{k=j}, X_{k-1=i1}, X_{k-2}, \ldots, X_0]P[X_{k-1=i1}, X_{k-2=i2}, \, , X_{0=ik}]$$

$$= P[X_{k=j}, X_{k-1=i1}]P[X_{k-1=i1}, X_{k-2}, \ldots, X_{0=ik}]$$

$$= P[X_{k=j}, X_{k-1=i1}]P[X_{k-1=i1}, X_{k-2=i2}, \ldots, X_0]P[X_{k-2=i2}, \ldots, X_0]$$

$$= P[X_{k=j}, X_{k-1=i1}]P[X_{k-1=i1}, X_{k-2=i2}], \ldots, P[X_{1=ik-1}, X_0] \quad ..(3.19)$$

$$P[X_{0=ik}] = p_{i1j}p_{i2i1}p_{i3i2} \cdots p_{ikik-1}P[X_{0=ik}]$$

Thus, when we know the initial state X0 and the state transition probabilities, we can evaluate the joint probability $P[X_k, X_{k-1}, \ldots, X_0]$.

3.5.1 State Transition Probability Matrix

The state transition probabilities as the entries of an $n \times n$ matrix P, where p_{ij} is the entry in the ith row and jth column,

P is called the *transition probability matrix*. It is a *stochastic matrix* because

for any row i, $\sum p_{ij} = 1$

$$P = \begin{pmatrix} p_{11} & p_{12} & p_{13} & \ldots & p_{1n} \\ p_{21} & p_{22} & p_{23} & \ldots & p_{2n} \\ \ldots & \ldots & \ldots & & \ldots \\ p_{n1} & p_{n2} & p_{n3} & & p_{nm} \end{pmatrix}$$

3.5.2 The *n*-Step State Transition Probability

Let $p_{ij}(n)$ denote the conditional probability that the process will be in state j after exactly n transitions, given that it is presently in state i. That is

$$p_{ij}(n) = P[X_{m+n} = j, X_m = i]$$

$$p_{ij}(0) = \begin{cases} 1 & i = j \\ 0 & i \neq j \end{cases}$$

$$p_{ij}(1) = p_{ij}$$

Consider the two-step transition probability $p_{ij}(2)$, which is defined by

$$p_{ij}(2) = P[X_{m+2} = j, X_m = i] \quad \ldots\ldots\ldots\ldots(3.20)$$

Where the second to the last equality is due to the Markov property. The final equation states that the probability of starting in state i and being in state j at the end of the second transition is the probability that we first go immediately from state i to an intermediate state k and then immediately from state k to state j; the summation is taken over all

48

possible intermediate states k. The following proposition deals with a class of equations called the *Chapman- Kolmogorov equations*, which provide a generalization of the preceding results obtained for the two-step transition probability[9].

CHAPTER- 4
METHODOLOGY AND MATHEMATICAL BACKGROUND

4.1 Methodology

The most important reliability indices are found namely failure rate (λ), repair rate (μ), MTTR, MTBF, MTTF Through data collection and analysis. An evaluation of Markov models used to obtain unit reliability and availability the operational data of these stations for period 2005 – 2010. The data of each year and for each unit is time scheduled. After tabulating all the data, for each unit the different type of failure taking into account the various sub units and systems were classified according to the classification Markov states were defined Failure rate ,repair rate, MTTR, MTTF, MTBF for each state were determined from the classified data. Subsequently availability and reliability were determined.

4.2 Mathematical Background

To model a unit or a system we use Markov model, which is a state-space representation. The matrix form of this representation is:

$$dP(t)/dt = P(t) A \quad \ldots\ldots\ldots\ldots\ldots(4.1)$$

Where $dP(t)/dt$ is a row-vector of derivative of the state probability consists of $dP1(t)/dt$, $dP2(t)/dt$.........; $P(t)$ is a row-vector consist of the elements of state probabilities; and A is the transitional rate matrix with elements $a_{ij}=X_{ij}$ for i#j, and $a_{ij}= \sum -\lambda\, ij$ for i j. The elements of each row of matrix A always add up to 0. The steady-state probability can be determine by much simpler task of solving the set of linear equations

$$PA = 0 \quad \ldots\ldots\ldots\ldots\ldots(4.2)$$

Solution of (4.2) requires an additional equation, which is obtained from the fact that the probability of states must always add up to 1; that is,

$$=0 \quad \ldots\ldots\ldots\ldots\ldots(4.3)$$

Substituting 3 in one equation of the set (4.2) and solve will gives steady state probabilities. The frequency of encountering each state equals the probability of been in the state times rate of departure from that state or rate of entry to that state. According to the definition

$$f_i = P_i \times \text{Rate of departure of state } P_i \quad \ldots\ldots(4.4)$$

4.3 Evaluating limiting state probabilities
4.3.1 Single repairable component

That stochastic transitional probability matrix ideally suited for the evaluation limiting state probabilities. The approach used to define α as a limiting state probability vector which remained unchanged when multiplied by the stochastic transitional probability matrix

$$\alpha P = \alpha \quad \ldots\ldots(4.5)$$

If α is given by [P_0 P_1] for the single repairable component, then

$$(1-\lambda dt)P_0 + \mu dt\, P_1 = P_0 \quad \ldots\ldots(4.6)$$

$$\lambda dt P_0 - (1- \mu dt)P_1 = P_1 \quad \ldots\ldots(4.7)$$

Where, $\quad P_0 = \dfrac{\lambda}{\lambda+\mu} \quad$ and $\quad P_1 = \dfrac{\mu}{\lambda+\mu}$

Rearranging Equation 1,2

$$(1-\lambda dt)P_0 + \mu dt\, P_1 = 0 \quad \ldots\ldots(4.8)$$

$$\lambda dt P_0 - (1- \mu dt)P_1 = 0 \quad \ldots\ldots(4.9)$$

In Equation 4.6 the value of dt, provided it is non-zero and finite, disappears to give

$$-\lambda P_0 + \mu P_1 = 0 \quad \ldots\ldots(4.10)$$

$$\lambda P_0 - \mu P_1 = 0 \quad \ldots\ldots(4.11)$$

It is convenient therefore to omit them entirely in formulating the initiating matrix and express the transition probabilities strictly in terms of the transition rates.

4.3.2 Two identical repairable components:

The state space diagram in shown in figure 3.4, the stochastic transitional probability matrix is

$$P = \begin{array}{c} 1 \\ 2 \\ 3 \end{array} \begin{bmatrix} 1-2\lambda & 2\lambda & 0 \\ \mu & 1-\lambda-\mu & \lambda \\ 0 & 2\mu & 1-2\mu \end{bmatrix}$$

Therefore, if limiting state probability vector is [P_1 P_2 P_3].Then

$$[P_1 \ P_2 \ P_3] \begin{bmatrix} 1-2\lambda & 2\lambda & 0 \\ \mu & 1-\lambda-\mu & \lambda \\ 0 & 2\mu & 1-2\mu \end{bmatrix} = [P_1 \ P_2 \ P_3]$$

Which is in explicit form, gives

$$P_1(1-2\lambda) + P_2\mu = P_1$$
$$P_1 2\lambda + P_2(1-\lambda-\mu) + P_3 2\mu = P_2$$
$$P_2\lambda + P_3(1-2\mu) = P_3$$

Rearranging gives

$$P_1(-2\lambda) + P_2\mu = 0$$
$$P_1 2\lambda - P_2(\lambda+\mu) + P_3 2\mu = 0$$
$$P_2\lambda - P_3(2\mu) = 0$$

And
$$P_1 + P_2 + P_3 = 1$$

The limiting state probabilities can be obtained by using matrix techniques, and are

$$P_1 = \underline{\qquad} \quad P_2 = \underline{\qquad} \quad P_3 = \underline{\qquad} \bigg\} \quad \ldots\ldots(4.12)$$

(a) Series connected components

In the case of two identical components connected in series, the up state of the system is state 1 and the down state is states 2 and 3, therefore

Availability, $A = P_1$

$$A = \frac{\mu 2}{\mu} \quad \ldots\ldots\ldots\ldots\ldots\ldots(4.13)$$

Unavailability, $U = P_2 + P_3$

$$\text{Unavailability}, U = \frac{2\lambda\mu+}{\mu} \quad \ldots\ldots\ldots\ldots\ldots\ldots(4.14)$$

(b) Parallel connected component

In the case of two identical components connected in Parallel, states 2 also becomes an up state giving

Availability, $A = P_1 + P_2$

Availability, $A = $ —— $\quad \ldots\ldots\ldots\ldots\ldots\ldots(4.15)$

Unavailability, $U = P_3$

Unavailability, $U = $ —— $\quad \ldots\ldots\ldots\ldots\ldots\ldots(4.16)$

It is to note that, in this syatem, each component is independent, and the expressions for P_1, P_2 and P_3 together with the values of availability and unavailability could have been obtained directly from the results of the single component system.

4.4 Evaluating time dependent state probability

The application of deferential equations method to more complicated systems and to verify the complexity that arises in attempting to derive general equations, consider the two identical component system shown in figure 3.4.

Let

$P_1(t)$ = probability that both component are in an operative state at time t

$P_2(t)$ = probability that one component is positive and one component is failed at time t, and

$P_3(t)$ = probability that both components are failed at time t.

Then the differential equations for this system are

Methodology and mathematical background

$$[dP_1/dt \quad dP_2/dt \quad dP_3/dt] = [P_1(t) \; P_2(t) \; P_3(t)] \begin{bmatrix} -2\lambda & 2\lambda & 0 \\ \mu & -(\lambda+\mu) & \lambda \\ 0 & 2\mu & -2\mu \end{bmatrix}$$

Assume that the system starts in state 1, then

$P_1(0) = 1, \quad P_2(0) = 1 \quad \text{and} \quad P_3(0) = 1$

The solution of above matrix is given as

$$\left. \begin{array}{l} P_1(t) = \dfrac{\mu 2}{\mu)} + \dfrac{2\lambda\mu+}{\mu)} \quad + \dfrac{}{} \\ \\ P_2(t) = \dfrac{}{\mu)} + \dfrac{}{\mu)} \quad - \dfrac{}{\mu)} \\ \\ P_3(t) = \dfrac{}{} \quad \dfrac{}{\mu)} \quad + \dfrac{}{} \end{array} \right\} \quad \ldots\ldots\ldots(4.17)$$

It can be seen that the derivation and the resulting general expression become rather than complex even in the case of two identical and repairable components.

4.5 Unit Modeling:

To model a hydro-unit generally according to its mode of operation. It can be divided into up-state and down-state. The state-space diagram is as follows:
The hydro-unit is transit from up-state to down-state, either due to forced or scheduled outages.
To derive the Markov model of a Hydro-unit we assume:
1. The failure and repair rates are exponentially distributed.
2. There is no transition between the scheduled and force outages. The unit after repairing is immediately returning to up-state.

From the above definition a developed Markov model is driven as follows:

Methodology and mathematical background

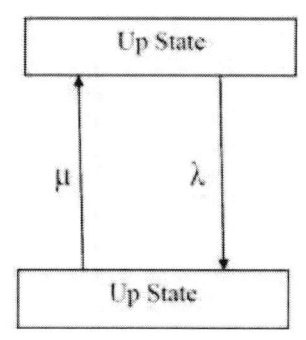

Fig4.1. Two State model

4.6 Three-State Model:

In most applications reported in the literature, which involve repairable components, reliability techniques are based on a two-state representation of components. In these models it is assumed that each component has an operating history made up of cycles of alternating periods spent in the "in-service" and "failed" states, as illustrated in Figure 1. The lines in the diagram indicate the possible transitions between states and each path is labeled with the mean frequency of that transition; these are the average failure frequency λ (the reciprocal of the mean in-service time) for transitions into the failure state and the reciprocal of the mean repair duration μ for the transitions back to normal. In most calculations, these mean values sufficiently describe the "up and down" cycles and the findings are independent of the actual up-time and down-time distributions[5].

The state of the entire system at any given moment is determined by the prevailing states of its components. Thus the system state will change if, and only if, there is a change in the state of any of its components. For example, a system that consists of two independent two-state components, i and j, may assume any of the states illustrated in Figure 2. In any given state, the system may be either successful or failed; one of the main tasks in reliability analysis is the division of system states into success and failure states. It is by no means obvious to which category a given state belongs; this can be established only after applying suitable criteria for system failure, which them selves are not always easy to define[10]

Most power system components go through the two-state cycles described in the previous section. The failure of such components, however, will often put the system through a more complicated routine than that discussed above. When a high-voltage device fails, for example, first the system protection will isolate a number of "healthy" components along with the faulted one; as soon as possible after that, all but the minimum number of components that must be kept out of service for the isolation of the failed device will be restored to operation through appropriate switching. From the system's point of view, therefore, the fault of such a device is followed by a system state where several components are out of service and, after switching, this is followed by another state where possibly only the faulted device is out. What actually occurs is that while the component is in the "failed" state, the system moves through two states, those "before switching" and "after switching." A model of this process can be constructed by considering such components to have three-state cycles consisting of an operating state, a state between the fault and switching (s state) and a "repair" state (r state) when the device is isolated for repair. Obviously, the system effects of the s and r states are very different. It should be noted that the r states, lasting until repairs are completed, are usually much longer in duration than the s states and, also, that there are only very weak restrictions as to the time distributions of any of the three states[5].

A system of two independent components i and j with three-state cycles will have a state transition diagram. The diagram is a composition of two "single" diagrams of the type shown in Figure, and is constructed in such a way as to allow for complete cycles of j starting from any of the states of i and vice versa. Once again, the diagram will assist in recognizing the various possible groups of failure states; it is just the assistance in sorting out the failure possibilities that makes this diagram useful.

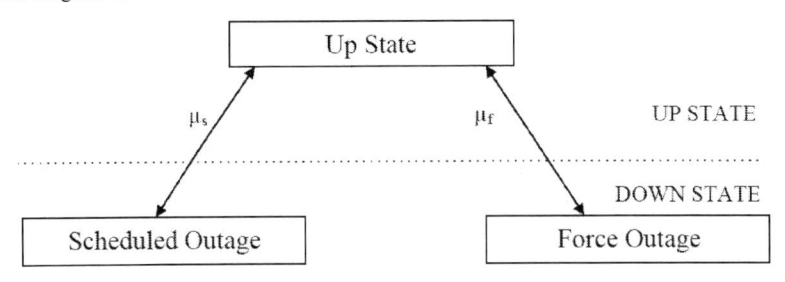

Fig4.2. Three state model

Methodology and mathematical background

The event of Hydro-unit and it's down states (Scheduled outage and Forced outage) into:

Scheduled outage:

State 1. Reserve, Preventive maintenance, and overhaul.

Forced outage:

State 2. Generator.

State 3. Turbine (inlet gate, penstock, spiral case, butter fly valve, turbine bearing, and runner).

State 4. Excitation system (thyristor, cooling system, equipped transformer, and etc...).

State 5. Governor system (servo motors, wicket gates, speed governor, and etc...).

State 6. Main Unit Transformer.

State 7. Main Unit Circuit Breaker.

State 8. External Effects.

More developed model is driven as follows:

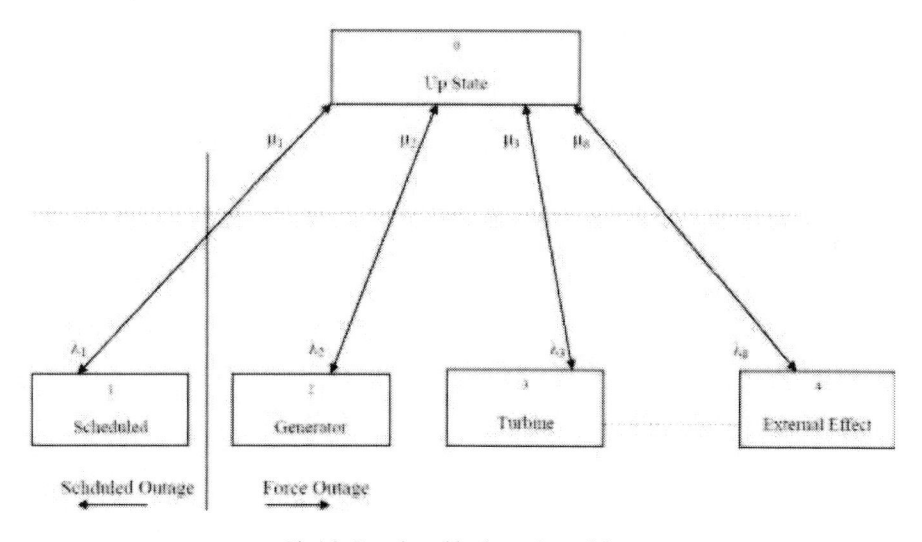

Fig4.3. Developed hydro-unit model

The state transition matrix of Fig. 3 is as follows:

$$P = \begin{pmatrix} -(\lambda_1+\lambda_2+\lambda_3+\lambda_4+\lambda_5+\ldots+\lambda_6) & \lambda_1 & \lambda_2 & \lambda_3 & \lambda_4 & \lambda_5 & \lambda_6 & \lambda_7 & \lambda_8 \\ \mu_1 & -\mu_1 & 0 & 0 & 0 & 0 & 0 & 0 & 0 \\ \mu_2 & -\mu_2 & 0 & 0 & 0 & 0 & 0 & 0 & 0 \\ \mu_3 & -\mu_3 & 0 & 0 & 0 & 0 & 0 & 0 & 0 \\ \mu_4 & -\mu_4 & 0 & 0 & 0 & 0 & 0 & 0 & 0 \\ \mu_5 & -\mu_5 & 0 & 0 & 0 & 0 & 0 & 0 & 0 \\ \mu_6 & -\mu_6 & 0 & 0 & 0 & 0 & 0 & 0 & 0 \\ \mu_7 & -\mu_7 & 0 & 0 & 0 & 0 & 0 & 0 & 0 \\ \mu_8 & -\mu_8 & 0 & 0 & 0 & 0 & 0 & 0 & 0 \end{pmatrix}$$

Table 4.1 State Probability Value

State No	State probability	
0	$\mu_1\mu_2\mu_3\mu_4\mu_5\mu_6\mu_7\mu_8 /D$	d_0/D
1	$\lambda_1\mu_2\mu_3\mu_4\mu_5\mu_6\mu_7\mu_8 /D$	d_1/D
2	$\mu_1\lambda_2\mu_3\mu_4\mu_5\mu_6\mu_7\mu_8 /D$	d_2/D
3	$\mu_1\mu_2\lambda_3\mu_4\mu_5\mu_6\mu_7\mu_8 /D$	d_3/D
4	$\mu_1\mu_2\mu_3\lambda_4\mu_5\mu_6\mu_7\mu_8 /D$	d_4/D
5	$\mu_1\mu_2\mu_3\mu_4\lambda_5\mu_6\mu_7\mu_8 /D$	d_5/D
6	$\mu_1\mu_2\mu_3\mu_4\mu_5\lambda_6\mu_7\mu_8 /D$	d_6/D
7	$\mu_1\mu_2\mu_3\mu_4\mu_5\mu_6\lambda_7\mu_8 /D$	d_7/D
8	$\mu_1\mu_2\mu_3\mu_4\mu_5\mu_6\mu_7\lambda_8 /D$	d_8/D

Where, $D = d_0 + d_1 + d_2 + d_3 + d_4 + d_5 + d_6 + d_7 + d$

Table 4.2 Frequency of Encountering States

Rate of departure	Frequency of state
$\lambda_1+\lambda_2+\lambda_3+\ldots+\lambda_8$	$(\lambda_1+\lambda_2+\lambda_3+\ldots+\lambda_8)\,d_0/D$
μ_1	$\mu_1 d_1/D$
μ_2	$\mu_2 d_2/D$
μ_3	$\mu_3 d_3/D$
μ_4	$\mu_4 d_4/D$
μ_5	$\mu_5 d_5/D$
μ_6	$\mu_6 d_6/D$
μ_7	$\mu_7 d_7/D$
μ_8	$\mu_8 d_8/D$

CHAPTER -5

PLANT MODELING AND ESTIMATION FOR PHPS

5.1 Introduction:

Pathri hydro power station (PHPS) has an installed capacity of 20.4MW.It consists of 3 identical independent unit of 6.8MW capacity per each. PHPS has been constructed on upper Ganga canal at 13 km downstream of holy city, Haridwar India. All the mechanical equipments were supplied by J.M.Vaith, Germany and electrical equipments by Siemens Germany. Each unit of PHPS units consists of several subunits such as Turbine, Generator, Excitation system, Speed Governor, Spiral case, etc. The maximum generation of 134.154 M.U. was achieved in the year of 1968-69 and minimum of 68.05 M.U. in 1977-78. The present generation is 90 M.U. per year. The Power House was constructed by P.W.D. (Irrigation branch) U.P. and taken over by the Hydel Deptt. later on. It remained under Distribution Wing of the then U.P.S.E.B. and finally handed over to Hydro Electric Projects, Dehradun in 1981. After trifurcation of UPSEB this Power House was handed over to U.P. Jal Vidut Nigam on 14-01-2000. From 09-11-2001 this Power House is running successfully under Uttarakhand Jal Vidut Nigam Ltd.

5.2 Unit Modelling:

To model a hydro-unit generally according to its mode of operation. It can be divided into up-state and down-state.

The hydro-unit is transit from up-state to down-state, either due to forced or scheduled outages.

To derive the Markov model of a Hydro-unit we assume:

1. The failure and repair rates are exponentially distributed.

2. There is no transition between the scheduled and force outages. The unit after repairing is immediately returning to up-state.

From the above definition a developed Markov model is driven as follows:

Plant modeling and estimation for PHPS

The event of Hydro-unit and it's down states (Scheduled outage and Forced outage) into:

Scheduled outage:

An outage that results when unit is deliberately taken out of service at selected time, usually for purpose of construction, preventive maintenance, repair or reserve.

State 1. Reserve, Preventive maintenance, and overhaul.

Forced outage:

An outage result from emergency conditions directly associated with component or unit requiring that unit be taken out of service immediately, either automatically or as soon as switching operations can be preformed.

State 2. Generator.

State 3. Turbine (inlet gate, penstock, spiral case, butter fly valve, turbine bearing, and runner).

State 4. Excitation system (thyristor, cooling system, equipped transformer, and etc...).

State 5. Governor system (servo motors, wicket gates, speed governor, and etc...).

State 6. Main Unit Transformer.

State 7. Main Unit Circuit Breaker.

State 8. External Effects.

5.3 Plant modeling:

To Model PHPS the three units should be studied together. The number of failure rates and repair rates of a unit for five year and for all the units are taken to determine the plant availability and reliability. The transition rate matrix of fig.3 is determined by the same way as the unit transition rate matrix. The state probabilities are determined by the same ways as for unit modeling. The probability of state 1 is the probability that the three units (PHPS) are up

$$P_1 = \mu_1\,\mu_2\,\mu_3 / \prod \mu$$

Probability of state 8 is the probability that all the units are down

$$P_8 = \lambda_1\,\lambda_2\,\lambda_3 / \prod \mu i$$

The frequency of encountering state 1 is

Plant modeling and estimation for PHPS

$$f_1 = (\lambda_1 + \lambda_2 + \lambda_3) P_1$$

The frequency of encountering state 8 is

$$f_8 = (\mu_1 + \mu_2 + \mu_3)/ P_8$$

5.4 State space diagram:

The number of states in the state space diagrams increases as the number of system components increase and as the number of state in which each system component can reside increases .If all states are represented. The number of states in the diagram is 2^3 for an 3-component system in which each component is represented as a 2-state model .The state space diagram all the system component are continuously operation either in series ,parallel or series/parallel. The important class of systems known as standby systems can also be modeled and analyzed using state space diagrams and Markov technique.

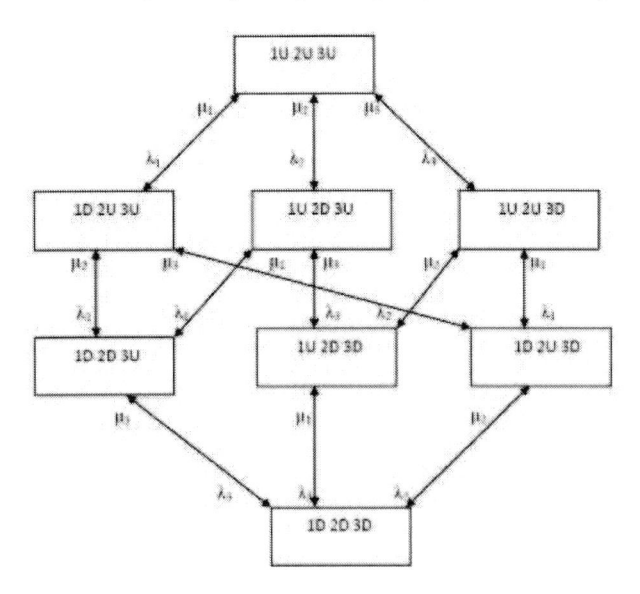

Fig5.1. State space diagram for PHPS

Plant modeling and estimation for PHPS

5.5 State probability, availability and reliability determination

The maximum number component of state in a three component, where each component can exist in two states, is 2^3 or 8. This is shown in fig.3 in λ and μ which represents the failure rate and repair rates of component and U and D indicates that the component is up or down respectively. This state space diagram can be modified by further knowledge of the system it is meant to represent, e.g., whether some states and transitions are inappropriate and whether derated states are also necessary. The state space diagram may under certain conditions be valid for a range of physical system. The states to be combined for system success and failure are

Table 5.1: PHPS [Unit-1] Failure Rates, Repair Rates State Probability and Availability, Reliability Determination 2005-10

State Number	Basic Event	No. of occurrence	Total repair times(hrs)	MTTR in hrs	MTBF in hrs	MTTF in hrs	Repair Rate in μ	Failure Rate in λ	Probability of occurrence	State Probability
0	Up State									0.995421
1	Reserve, preventive maintenance and overhaul									
2	Generator									
3	Turbine(inlet gate, penstock, spiral case, butter fly valve, turbine bearing, and runner)	11	54.2	4.9272727	3986.182	3981.255	0.202952	0.00025	0.00123608	0.0012293
4	Excitation system(thyrister, cooling system, equipped transformer, and etc.)	4	9.55	2.3875	10962	10959.61	0.418848	0.00009	0.000217798	0.0002144
6	Main unit transformer	7	26.55	3.7928571	6264	6260.207	0.263653	0.00015	0.000605501	0.0006056
7	Main unit circuitbreaker									
8	External Effect									
			108.22				0.885454	0.00050	0.002059387	
		Reliability=0.996650			Availability=0.997940					

Plant modeling and estimation for PHPS

Table 5.2: PHPS [Unit-2] Failure Rates, Repair Rates State Probability and Availability, Reliability Determination 2005-10

Down States event of hydro unit 2

State	Basic Event	No.of occurrence	Total repair times(hrs)	MTTR in hrs	MTBF in hrs	MTTF in hrs	Repair Rate in μ	Failure Rate in λ	Probability of occurrence	State Probability
0	Up State									0.564728
1	Reserve,preventive maintenance and overhaul									
2	Generator	4	31.6	7.9	10962	10954.1	0.126582	0.00009	0.000721191	0.0038339
3	Turbine(inlet gate,penstock,spiral case,butter fly valve,turbine bearing,and runner)	3	214.25	71.416667	14616	14544.58	0.014000	0.00007	0.00491019	0.028115
4	Excitation system(thyrister,cooling system,equipped transformer,and etc.)									
5	Governor system(servo motor,wicket gates,speed governor,and etc)									
6	Main unit transformer	8	267.3	33.4125	5481	5447.588	0.029929	0.00018	0.006133449	0.4063897
7	Main unit circuitbreaker									
8	External Effect	3	36	12	14616	14604	0.083333	0.00007	0.000821693	0.00038339
			549.15				0.253847	0.00041	0.012586523	
			Reliability=0.971118		Availability=0.9874134					

Table 5.3: PHPS [Unit-3] Failure Rates, Repair Rates State Probability and Availability, Reliability Determination 2005-10

Down States event of hydro unit 3

State	Basic Event	No.of occurrence	Total repair times(hrs)	MTTR in hrs	MTBF in hrs	MTTF in hrs	Repair Rate in μ	Failure Rate in λ	Probability of occurrence	State Probability
0	Up State									0.9830485
1	Reserve,preventive maintenance and overhaul									
2	Generator									
3	Turbine(inlet gate,penstock,spiral case,butter fly valve,turbine bearing,and runner)	16	119.25	7.453125	2740.5	2733.047	0.134172	0.00037	0.00273	0.00273785
4	Excitation system(thyrister,cooling system,equipped transformer,and etc.)	3	4.2	1.4	14616	14614.6	0.714286	0.00007	0.00010	9.6432E-06
5	Governor system(servo motor,wicket gates,speed governor,and etc)	4	151.15	37.7875	10962	10924.21	0.026464	0.00001	0.00346	0.00033151
6	Main unit transformer									
7	Main unit circuitbreaker									
8	External Effect	3	32.12	10.706667	14616	14605.29	0.0934	0.00007	0.00073	0.00074211
			306.72				0.968321	0.00051	0.00701	
			Reliability=0.985786		Availability=0.99298503					

Plant modeling and estimation for PHPS

Table 5.4: PHPS State Probability and Availability, Reliability Determination 2005-10

State Number	State Probability	Frequency Of State
1	0.335310000	0.000146540
2	0.213010000	0.000056000
3	0.203061000	0.000042600
4	0.211300000	0.000000050
5	0.000000094	0.000000011
6	0.000000096	0.000000117
7	0.000000003	0.000000006
8	0.000000000	0.000000000

Series system

success = state 1

Failure = states 2,3,4,5,6,7,8

Parallel redundant system

success = states 1,2,3,4,5,6,7

Failure = states 8

2-out-of-3 system

success = states 1,2,3,4

Failure = states 5,6,7,8

Availability of PHPS = 0.97012

2-out-of 3 system, State Probability = 0.962681

So, Reliability of PHPS = 0.962681

CHAPTER- 6

PLANT MODELLING AND ESTIMATION FOR CHPS

6.1 Introduction

Chilla hydro power station (CHPS) has an installed capacity of 144MW. It consists of 4 identical independent units of 36 MW capacities per each. CHPS is a runoff river scheme constructed under Garhwal Rishikesh Chilla hydel scheme in the river Ganga. It comprises a diversion barrage across the river Ganga at Pashulok 5 km downstream of Rishikesh town. Each unit of CHPS comprises vertical shaft Kaplan turbine of rated head 32.5 meter. There are separate penstocks for each unit. The length of power canal and tailrace canal are respectively 14.3km and 1.2 km.

6.2 Unit modelling

To model a hydro-unit generally according to its mode of operation. It can be divided into up-state and down-state.

The hydro-unit is transit from up-state to down-state, either due to forced or scheduled outages.

To derive the Markov model of a Hydro-unit we assume:

1. The failure and repair rates are exponentially distributed.
2. There is no transition between the scheduled and force outages. The unit after repairing is immediately returning to up-state.

From the above definition a developed Markov model is driven as follows:

The event of Hydro-unit and it's down states (Scheduled outage and Forced outage) into:

Scheduled outage:

An outage that results when unit is deliberately taken out of service at selected time, usually for purpose of construction, preventive maintenance, repair or reserve.

State 1. Reserve, Preventive maintenance, and overhaul.

Forced outage:

An outage result from emergency conditions directly associated with component or unit requiring that unit be taken out of service immediately, either automatically or as soon as switching operations can be preformed.

State 2. Generator.

State 3. Turbine (inlet gate, penstock, spiral case, butter fly valve, turbine bearing, and runner).

State 4. Excitation system (thyristor, cooling system, equipped transformer, and etc...).

State 5. Governor system (servo motors, wicket gates, speed governor, and etc...).

State 6. Main Unit Transformer.

State 7. Main Unit Circuit Breaker.

State 8. External Effects.

6.3 Plant Modeling:

To Model CHPS the number of failure rates and repair rates of a unit for five year and for all the four units are taken to determine the plant availability and reliability. The transition rate matrix of fig.4 is determined by the same way as the unit transition rate matrix. The probability of state 1 is the probability that the four units (CHPS) are up

$$P_1 = \mu_1 \mu_2 \mu_3 \mu_4 / \prod$$

Probability of state 16 is the probability that all the units are down

$$P_{16} = \lambda_1 \lambda_2 \lambda_3 \lambda_4 / \prod \qquad \mu i$$

The frequency of encountering state 1 is

$$f_1 = (\lambda_1 + \lambda_2 + \lambda_3 + \lambda_4) P_1$$

The frequency of encountering state 8 is

$$F_{16} = (\mu_1 + \mu_2 + \mu_3 + \mu_4) / P_{16}$$

6.4 State Space Diagram:

The number of states in the state space diagrams increases as the number of system components increase and as the number of state in which each system component can reside increases .If all states are represented. The number of states in the diagram is 2^4 for an 4-component system in which each component is represented as a 2-state model .The state space diagram all the system component are continuously operation either in series ,parallel or series/parallel. The important class of systems known as standby systems can also be modeled and analyzed using state space diagrams and Markov technique.

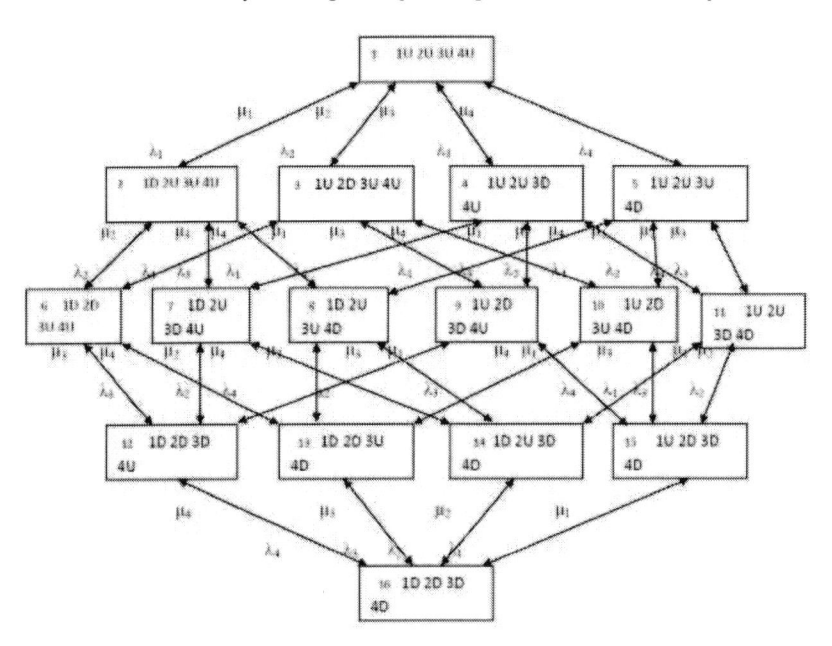

Fig6.1.State space diagram for CHPS

Plant modelling and estimation for CHPS

6.5 State Probability, Availability and Reliability Determination

The solution of continuous or discrete Markov processes it is desirable first to construct the appropriate state space diagram and insert the relevant transition rates. All relevant State in which the system can reside should be including in such a diagram and all known ways in which transition between states can occur should be inserted. The relevant state space diagram for the simple single component is shown in fig 7.1. In the case of continuous Markov processes they are usually represented by a transition rate as shown in above figure, by the transition λ and μ from the operating and failed states respectively. The maximum number component of state in a three component, where each component can exist in two states, is 2^4 or 16. This is shown in fig.3 in λ and μ which represents the failure rate and repair rates of component and U and D indicates that the component is up or down respectively. This state space diagram can be modified by further knowledge of the system it is meant to represent,e.g.,whether some states and transitions are inappropriate and whether derated states are also necessary.

Table 6.1: CHPS [Unit-1] Failure Rates, Repair Rates State Probability and Availability, Reliability Determination 2005-10

StateNo	Basic Event	No.of occurrence	Total repair times(hrs)	MTTR in hrs	MTBF in hrs	MTTF in hrs	Repair Rate in μ	Failure Rate in λ	Probability of occurrence	State Probability
0	Up State									0.9754686
1	Reserve,preventive maintenance and overhaul									
2	Generator	3	3.3	1.1	13080.66667	13079.57	0.909090	0.000076	0.00008	0.00008
3	Turbine(inlet gate,penstock,spiral case,butter fly valve,turbine bearing,and runner)	3	11.02	3.673333	13080.66667	13076.99	0.272232	0.000076	0.00028	0.000281
4	Excitation system(thyrister,cooling system,equipped transformer,and etc.)	1	4.05	4.05	39242	39237.95	0.246914	0.000025	0.00010	0.0001025
5	Governor system(servo motor,wicket gates,speed governor,and etc)	1	0.45	0.45	39242	39241.55	2.222222	0.000025	0.00001	0.00001
6	Main unit transformer									
7	Main unit circuitbreaker									
8	External Effect	3	2.05	0.683333	13080.66667	13079.98	1.463415	0.000076	0.00005	0.00005
			20.87				5.113874	0.000280	0.000511	
	Availability=0.99943668					Reliability=0.975750198				

69

Plant modelling and estimation for CHPS

Table 6.2: CHPS [Unit-2] Failure Rates, Repair Rates State Probability and Availability, Reliability Determination 2005-10

StateNo	Basic Event	No.of occurrence	Total repair times(hrs)	MTTR in hrs	MTBF in hrs	MTTF in hrs	Repair Rate in µ	Failure Rate in λ	Probability of occurrence	State Probability
0	Up State									0.96845838
1	Reserve,preventive maintenance and overhaul									
2	Generator	6	15.08	2.513333	6540.333333	6537.82	0.397878	0.00015296	0.00038	0.00038369
3	Turbine(inlet gate,penstock,spiral case,butter fly valve,turbine bearing,and runner)	4	15.55	3.8875	9810.5	9806.613	0.257235	0.00010197	0.00040	0.00039548
4	Excitation system(thyrister,cooling system,equipped transformer,and etc.)	4	10.1	2.525	9810.5	9807.975	0.39604	0.00010196	0.00026	0.00025672
5	Governor system(servo motor,wicket gates,speed governor,and etc)	1	3.4	3.4	39242	39238.6	0.294118	0.00003	0.00009	0.00010142
6	Main unit transformer									
7	Main unit circuitbreaker									
8	External Effect	3	15.2	5.066667	13080.66667	13075.6	0.197368	0.00008	0.00039	0.00040439
			59.33				1.542638	0.00046	0.00151	
			Availability=0.998817854				Reliability=0.968853863			

Table 6.3: CHPS [Unit-3] Failure Rates, Repair Rates State Probability and Availability, Reliability Determination 2005-10

StateNo	Basic Event	No.of occurrence	Total repair times(hrs)	MTTR in hrs	MTBF in hrs	MTTF in hrs	Repair Rate in µ	Failure Rate in λ	Probability of occurrence	State Probability
0	Up State									0.97884441
1	Reserve,preventive maintenance and overhaul									
2	Generator	1	1.1	1.1	39242	39240.9	0.90909	0.00003	0.00003	0.00003
3	Turbine(inlet gate,penstock,spiral case,butter fly valve,turbine bearing,and runner)	7	28.35	4.05	5606	5601.95	0.246914	0.00018	0.00072	0.00072806
4	Excitation system(thyrister,cooling system,equipped transformer,and etc.)	1	5.05	5.05	39242	39236.95	0.19802	0.00003	0.00013	0.00015126
5	Governor system(servo motor,wicket gates,speed governor,and etc)	1	1.15	1.15	39242	39240.85	0.869565	0.00003	0.00003	0.00003
6	Main unit transformer									
7	Main unit circuitbreaker									
8	External Effect	2	8.35	4.175	19621	19616.83	0.239521	0.00005	0.00021	0.00020845
			44				2.46311	0.00031	0.00112	
			Availability=0.999563026				Reliability=0.979572967			

Plant modelling and estimation for CHPS

Table 6.4: PHPS [Unit-1] Failure Rates, Repair Rates State Probability and Availability, Reliability Determination 2005-10

Down States event of hydro unit 4

StateNo	Basic Event	No.of occurrence	Total repair times(hrs)	MTTR in hrs	MTBF in hrs	MTTF in hrs	Repair Rate in µ	Failure Rate in λ	Probability of occurrence	State Probability
0	Up State									0.99490941
1	Reserve, preventive maintenance and overhaul									
2	Generator									
3	Turbine(inlet gate, penstock, spiral case, butter fly valve, turbine bearing, and runner)	1	1	1	39242	39241	1	0.00003	0.00003	0.00003
4	Excitation system(thyrister, cooling system, equipped transformer, and etc.)	1	2.34	2.34	39242	39239.66	0.42735	0.00003	0.00006	0.00007
5	Governor system(servo motor, wicket gates, speed governor, and etc)									
6	Main unit transformer									
7	Main unit circuitbreaker									
8	External Effect	1	13.25	13.25	39242	39228.75	0.075472	0.00003	0.00034	0.00039728
			16.59					0.00008	0.00042	
			Availability=0.999503465				Reliability=0.99530669			

Table 6.5: CHPS State Probabilities and Availability, Reliability Determination 2005-10

State Number	State Probability	Frequency Of State
1	0.20162000000	0.0001350600000
2	0.10100000000	0.0000302450000
3	0.10400000000	0.0000101580000
4	0.09150000000	0.0000002035600
5	0.13010000940	0.0000000158000
6	0.10030009600	0.0000001430000
7	0.10146000300	0.0000000033000
8	0.10526000000	0.0000000042300
9	0.00913200000	0.0000000002350
10	0.00104500000	0.0000000001145
11	0.00135000000	0.0000000000253
12	0.00231000000	0.0000000000142
13	0.00505100000	0.0000000000113
14	0.00006023000	0.0000000000052
15	0.00001032000	0.0000000000032
16	0.00000000021	0.0000000000001

Plant modelling and estimation for CHPS

The state space diagram may under certain conditions be valid for a range of physical system. The states to be combined for system success and failure are

    2-out-of-4 system

$$\text{Success} = \text{states } 1,2,3,4,5,6,7,8,9,10,11$$
$$\text{Failure} = \text{states } 12, 13,14,15,16$$

    3-out-of-4 system

$$\text{Success} = \text{states } 1, 2,3,4,5$$
$$\text{Failure} = \text{states } 6,7,8,9,10,11,12,13,14,15,16$$

Then, 2-out-of 4 systems

$$\text{State Probability} = 0.951120$$

    3-out-of 4 system   State Probability = 0.007377

So, Reliability of CHPS = 0.951120

Availability of CHPS = 0.960530

CHAPTER 7

RESULTS, DISCUSSION AND CONCLUSION

7.1 Unit wise Reliability and Availability:

Unit wise reliability and availability of CHPS and PHPS is given below.

According to the operational behavior of the unit through five year its model is derived in chapter 5 for PHPS. The Reliability and Availability each unit of PHPS is given in Table 7.1.

Table 7.1 System Availability and Reliability of PHPS 2005-10

UNIT	Availability	Reliability
1	0.9979	0.9967
2	0.9874	0.9711
3	0.9930	0.9858

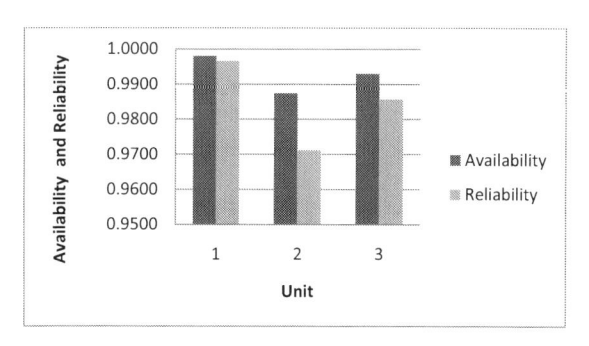

Fig.7.1. Units Availability & Reliability histogram for PHPS 2005-2010

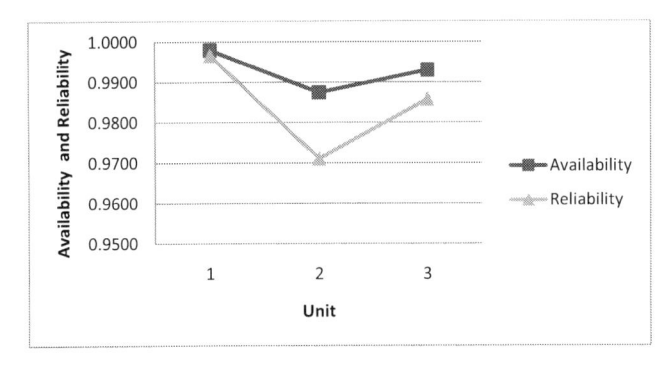

Fig.7.2. Units Availability and Reliability curves for PHPS 2005-2010

The Reliability and Availability each unit of CHPS is given in Table 7.2. According to the operational behavior of the unit through five year its model is derived in chapter 6 for CHPS.

Table 7.2 System Availability and Reliability of CHPS 2005-10

UNIT	Availability	Reliability
1	0.9994	0.9758
2	0.9988	0.9689
3	0.9996	0.9796
4	0.9995	0.9953

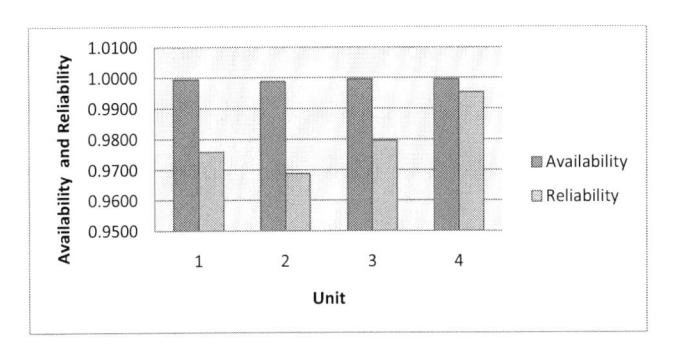

Fig.7.3 Units Availability & Reliability histogram for CHPS 2005-2010

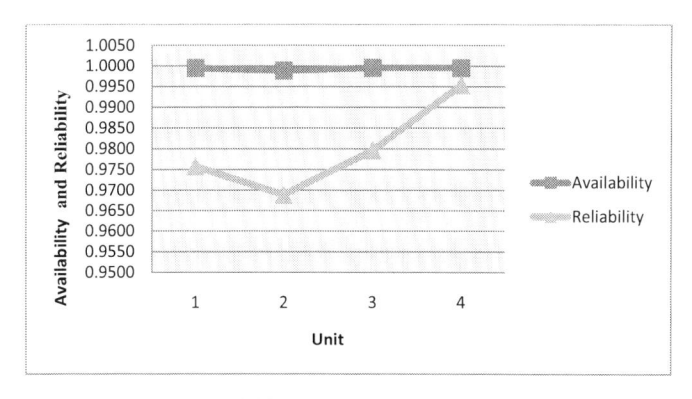

Fig.7.4. Units Availability and Reliability curves for 2005-2010

7.2 Reliability and Availability from Plant Modelling

The reliability and availability of PHPS and CHPS, According to the operational behavior of the unit through five year its model is derived in chapter 5 and chapter 6.

(a) *For PATHRI Hydropower station:*

Availability of PHPS is given bellow, since all the three units are operated in parallel. 2-out-3 availability is evaluated also for determination of reliability.The average generation for 2005-2010 was adequate and sufficient of demand, such that 2 or more units are operating. According to the definition of reliability the reliability of system will be probability 2-out-3.The intended condition is the probability of 2 out of 3 because under this condition the generation is sufficient and the power plant is reliable.

From Table: 6.1

2-out-of-3 system   success = states 1,2,3,4

Failure = states 5,6,7,8

2-out-of 3 system, State Probability = 0.962681

Then,   Relaibility of PHPS R_{PHPS} = 0.962681

And   Availability of PHPS $A_{PHPS} = 0.97012$

(b) For CHILLA Hydropower station:

Availability of CHPS is given bellow, since all the four units are operated in parallel. 2-out-4 availability is evaluated also for determination of reliability. The average generation for 2005-2010 was adequate and sufficient of demand, such that 2 or more units are operating. According to the definition of reliability the reliability of system will be probability 2-out-4. The intended condition is the probability of 2 out of 4 because under this condition the generation is sufficient and the power plant is reliable.

From Table

2-out-of-4 system

Success = states 1,2,3,4,5,6,7,8,9,10,11

Failure = states 12, 13,14,15,16

3-out-of-4 system

Success = states 1, 2,3,4,5

Failure = states 6,7,8,9,10,11,12,13,14,15,16

Then,   2-out-of 4 systems   State Probability = 0.951120

3-out-of 4 system   State Probability = 0.007377

Then,   Reliability of CHPS = R_{CHPS} = 0.951120

And   Availability of CHPS = A_{CHPS} = 0.960530

7.3 Discussion and Conclusion

- The weak points that cause poor point reliability and availability for PHPS and CHPS are given in Table 7.1 and 7.2.

Table 7.3: Unit Major Faults That Affect the Reliability Indices for PHPS

Unit No.	Cause of Fault	Down during 39242 hours of operation Fault
Unit-1	Turbine (inlet gate, penstock,…etc)	54.2
Unit-2	Main Unit Transformer	267.3
Unit-2	Turbine (inlet gate, penstock,…etc)	214.25
Unit-3	Turbine (inlet gate, penstock,…etc)	119.25
Unit-3	Governor system (servo motors, wicket gates, Speed governor and etc...).	151.15

Table 7.4: Unit Major Faults That Affect the Reliability Indices for CHPS

Unit No.	Cause of fault	Down during 39242 hours of operation Fault
Unit-1	Turbine (inlet gate, penstock,…etc)	11.02
Unit-2	Turbine (inlet gate, penstock,…etc)	15.55
Unit-3	Turbine (inlet gate, penstock,…etc)	28.35
Unit-3	Main Unit Transformer	8.25

- Main unit transformer is highly reliable and efficient equipment used for bulk transfer of power from one voltage level to another and is always under the influence of electrical, mechanical, thermal and environmental stresses which cause the degradation of insulation quality and the ultimate failure of a transformer leading to major breakdown of the power system itself.

- Some of the fault like partial discharge, overheating, winding circulating currents, arcing and continuous sparking can cause the deterioration of the insulation. Transformer insulation condition is the basic indicator of the best operation of the transformer. The transformer insulation is affected by aging, transient voltage and high operating temperature.
- The availability of the machines at CHPS and PHPS are of the order of 0.97012 and 0.960530 respectively which is of a very high order. This can be attributed to the fact that these power stations are canal based and therefore the abrasion erosion of the underwater parts of the hydro turbine is not substantial as the silt load is within the permissible limit of 3000ppm.
- The study of the plant availability and reliability that the maintenance program and skill of Engineers and technicians play an important role for improving the performance of the units and increasing the availability and reliability of the units and the power plant.

APPENDIX 1.1

BREAK DOWN RECORD OF CHILLA HYDRO POWER STATION (CHPS)
Unit-1

ROTOR EARTH FAULT

S.No.	Duration				Outage Hours
	From		To		
	Date	Time in Hours	Date	Time in Hours	
1	6/9/2005	10:27	6/9/2005	11:30	1:03
2	6/10/2005	5:00	6/10/2005	5:40	0:40
3	12/7/2006	5:50	12/7/2006	7:10	1:20
			Total outage hours:		3:03

EARTH FAULT IN UAT

S.No.	Duration				Outage Hours
	From		To		
	Date	Time in Hours	Date	Time in Hours	
1	22/09/2007	17:10	22/09/2007	18:00	0:50
2	23/09/2007	8:45	23/09/2007	9:05	0:20
			Total outage hours		1:10

Z CLAMP FAILURE

S.No.	Duration				Outage Hours
	From		To		
	Date	Time in Hours	Date	Time in Hours	
1	21/09/2005	9:10	21/09/2005	10:05	0:55
			Total outage hours		0:55

GUIDE VANE FAILURE

S.No.	Duration				Outage Hours
	From		To		
	Date	Time in Hours	Date	Time in Hours	
1	14/12/2007	6:38	14/12/2007	12:40	6:02
	Total outage hours				6:02

THE UNIT TRIPPED ON MECHANICAL FAULT

S.No.		Duration			Outage Hours
		From	To		
	Date	Time in Hours	Date	Time in Hours	
1	21/08/2008	9:45	21/08/2008	10:05	0:20
	Toatal outage hours				0:20

GLAND SEAL FAILURE

S.No.	Duration				Outage Hours
	From		To		
	Date	Time in Hours	Date	Time in Hours	
1	10/10/2008	4:40	10/10/2008	8:45	4:05
	Total outage hours				4:05

APPENDIX 1.2

BREAK DOWN RECORD OF CHILLA HYDRO POWER STATION (CHPS)
Unit-2

TURBINE GATE BEARING FAILURE (TGB)

S.No.	Duration				Outage Hours
	From		To		
	Date	Time in Hours	Date	Time in Hours	
1	27/08/2005	10:20	27/08/2005	15:05	4:45
2	31/08/2005	6:00	31/08/2005	8:20	2:20
			Total outage hours		7:05

GLAND SEAL LEAKAGE

S.No.	Duration				Total Outage hours
	From		To		
	Date	Time in Hours	Date	Time in Hours	
1	26/07/2006	3:45	26/07/2006	5:30	1:45
2	5/7/2008	2:45	5/7/2008	9:50	7:05
			Total outage hours		8:50

BURSTING OF CT

S.No.	Duration				Outage Hours
	From		To		
	Date	Time in Hours	Date	Time in Hours	
1	2/9/2006	11:00	2/9/2006	20:55	9:55
			Total outage hours		9:55

PROBLEM ON ROTATING SLEEVE

S.No.	Duration				Outage Hours
	From		To		
	Date	Time in Hours	Date	Time in Hours	
1	12/8/2007	7:05	12/8/2007	10:45	3:40
	Total outage hours				3:40

LEAKAGE FROM TOP COVER

S.No.	Duration				Outage Hours
	From		To		
	Date	Time in Hours	Date	Time in Hours	
1	1/11/2007	3:25	1/11/2007	6:40	3:15
	Total outage hours				3:15

MECHANICAL FAULT

S.No.	Duration				Outage Hours
	From		To		
	Date	Time in Hours	Date	Time in Hours	
1	18/04/2008	6:10	18/04/2008	8:25	2:15
2	21/08/2008	2:20	21/08/2008	2:45	0:25
	Total outage hours				2:40

EXCITATION PROBLEM

S.No.	Duration				Outage Hours
	From		To		
	Date	Time in Hours	Date	Time in Hours	
1	14/07/2008	5:25	14/07/2008	6:10	0:45
2	6/10/2008	0:35	6/10/2008	2:20	1:45
3	7/10/2008	12:45	7/10/2008	13:15	0:30
4	1/10/2008	7:10	1/10/2008	14:20	7:10
	Total outage hours				10:10

ELECTRICAL FAULT

S.No.	Duration				Outage Hours
	From		To		
	Date	Time in Hours	Date	Time in Hours	
1	17/08/2008	4:58	17/08/2008	5:15	0:17
2	6/1/2009	1:10	6/1/2009	2:10	1:00
	Total outage hours				1:17

ROTOR EARTH FAULT

S.No.	Duration				Outage Hours
	From		To		
	Date	Time in Hours	Date	Time in Hours	
1	24/08/2008	2:47	24/08/2008	3:35	0:48
2	25/08/2008	7:15	25/08/2008	8:45	1:30
	Total outage hours				2:18

LEAKAGE FROM OIL HEADER

S.No.	Duration				Outage Hours
	From		To		
	Date	Time in Hours	Date	Time in Hours	
1	9/10/2008	4:20	9/10/2008	6:30	2:10
	Total outage hours				2:10

APPENDIX 1.3

BREAK DOWN RECORD OF CHILLA HYDRO POWER STATION (CHPS)
Unit-3

TURBINE GATE BEARING (TGB) FAILURE

S.No.	Duration				Outage Hours
	From		To		
	Date	Time in Hours	Date	Time in Hours	
1	27/08/2005	1:05	27/08/2005	13:05	12:00
2	9/9/2005	1:50	9/9/2005	5:35	3:45
3	14/09/2005	16:35	14/09/2005	17:45	1:10
4	15/09/2005	22:50	15/09/2005	23:30	0:40
5	31/07/2008	0:45	31/07/2008	2:15	1:30
			Total outage hours		19:05

GLAND SEAL FAILURE

S.No.	Duration				Outage Hours
	From		To		
	Date	Time in Hours	Date	Time in Hours	
1	27/08/2005	3:55	27/08/2005	12:00	8:05
2	7/2/2008	17:25	7/2/2008	18:50	1:25
			Total outage hours		9:30

GOVERNOR FAILURE

S.No.	Duration				Outage Hours
	From		To		
	Date	Time in Hours	Date	Time in Hours	
1	8/10/2007	9:25	8/10/2007	10:40	1:15
			Total outage hours		1:15

FLASH OVER IN EXCITATION PROBLEM

S.No.	Duration				Outage Hours
	From		To		
	Date	Time in Hours	Date	Time in Hours	
1	3/4/2008	4:20	3/4/2008	9:25	5:05
			Total outage hours		5:05

OIL HEADER PROBLEM

S.No.	Duration				Outage Hours
	From		To		
	Date	Time in Hours	Date	Time in Hours	
1	22/07/2008	7:13	22/07/2008	7:13	0:40
			Total outage hours		0:40

APPENDIX 1.4

BREAK DOWN RECORD OF CHILLA HYDRO POWER STATION (CHPS)
Unit-4

Turbine Gate Bearing Failure (TGB)

S.No.	Duration				Outage Hours
	From		To		
	Date	Time in Hours	Date	Time in Hours	
1	6/10/2007	22:24	6/10/2007	23:24	1:00

OIL LEAKAGE FROM RUNNER BLADE

S.No.	Duration				Outage Hours
	From		To		
	Date	Time in Hours	Date	Time in Hours	
1	22/03/2008	3:58	22/03/2008	17:23	13:25

THRUST BEARING COOLER FAILURE

S.No.	Duration				Outage Hours
	From		To		
	Date	Time in Hours	Date	Time in Hours	
1	16/08/2008	17:31	16/08/2008	20:05	2:34

APPENDIX 2.1
BREAK DOWN RECORD OF PATHRI HYDRO POWER STATION (PHPS)
Unit-1

PILOT VALVE FAILURE

S.No.	Duration				Outage Hours
	From		To		
	Date	Time in Hours	Date	Time in Hours	
1	21/02/2007	3:10	21/02/2007	10:00	6:50
2	14/05/2007	10:15	14/05/2007	12:45	2:30
3	28/05/2007	14:30	28/05/2007	20:00	5:30
4	13/08/2008	16:20	13/08/2008	22:30	6:10
5	30/03/2010	12:40	30/03/2010	13:50	1:10
			Total outage hours		22:10

SLIP RING FAILURE

S.No.	Duration				Outage Hours
	From		To		
	Date	Time in Hours	Date	Time in Hours	
1	13/04/2004	20:00	13/04/2004	21:45	1:45
2	27/06/2006	8:00	28/06/2006	17:00	33:00:00
3	23/02/2006	9:15	26/02/2006	17:00	79:45:00
			Total outage hours		18:30

FAILURE OF BRUSH BEARING

S.No.	Duration				Outage Hours
	From		To		
	Date	Time in Hours	Date	Time in Hours	
1	13/09/2005	9:30	13/09/2005	11:55	0:40
			Total outage hours		0:40

COOLER FAILURE

S.No.	Duration				Outage Hours
	From		To		
	Date	Time in Hours	Date	Time in Hours	
1	21/09/2005	9:30	21/09/2005	11:15	1:45
			Total outage hours		1:45

CLEANING OF THRUST PILOT AND SOME TANK

S.No.	Duration				Outage Hours
	From		To		
	Date	Time in Hours	Date	Time in Hours	
1	27/09/2006	17:45	28/09/2006	4:40	10:55
			Total outage hours		10:55

FAILURE OF AUXILIRY ROTOR

S.No.	Duration				Outage Hours
	From		To		
	Date	Time in Hours	Date	Time in Hours	
1	15/11/2006	10:45	16/11/2006	5:20	19:15
			Total outage hours		19:15

PROBLEM ON CARBON BRUSH

S.No.	Duration				Outage Hours
	From		To		
	Date	Time in Hours	Date	Time in Hours	
1	26/01/2007	12:05	26/01/2007	12:30	0:25
			Total outage hours		0:25

PROBLEM ON UGB COOLER

S.No.	Duration				Outage Hours
	From		To		
	Date	Time in Hours	Date	Time in Hours	
1	28/08/2007	18:30	29/08/2007	6:10	11:40
2	9/9/2007	15:05	9/9/2007	16:05	1:00
3	31/08/2008	8:05	31/08/2008	9:05	1:00
			Total outage hours		13:40

DIFFERENTIAL FAILURE

S.No.	Duration				Outage Hours
	From		To		
	Date	Time in Hours	Date	Time in Hours	
1	23/06/2008	17:00	25/06/2008	19:15	50:15:00
2	26/06/2008	10:15	13/07/2008	14:45	412:30:00
3	15/07/2008	16:45	15/078/2008	20:45	4:00
			Total outage hours		466:45:00

GOVERNOR FAILURE

S.No.	Duration				Outage Hours
	From		To		
	Date	Time in Hours	Date	Time in Hours	
1	13/08/2008	16:20	13/08/2008	22:30	6:20
			Total outage hours		6:20

FAILURE OF GOVERNOR FILTER

S.No.	Duration				Outage Hours
	From		To		
	Date	Time in Hours	Date	Time in Hours	
1	25/08/2008	8:15	25/08/2008	9:50	1:35
			Total outage hours		1:35

FAILURE OF MACHINE(MAIN EXCITER ,PILOT EXCITER.PENDULAM MOTER AND REGULATOR RING)

S.No.	Duration				Outage Hours
	From		To		
	Date	Time in Hours	Date	Time in Hours	
1	25/03/2006	10:00	19/04/2006	15:30	605:30:00
2	27/09/2006	17:45	28/09/2006	4:40	10:55
3	28/05/2007	14:30	28/05/2007	20:00	5:50
4	28/08/2007	18:30	29/08/2007	6:10	11:40
			Total outage hours		9:55

APPENDIX 2.2
BREAK DOWN RECORD OF PATHRI HYDRO POWER STATION (PHPS)
Unit-2

MAIN TRANSFORMER FAILURE

S.No.	Duration				Outage Hours
	From		To		
	Date	Time in Hours	Date	Time in Hours	
1	24/09/2005	22:10	24/09/2005	22:30	0:20
2	9/3/2006	5:30	12/3/2006	21:50	88:20:00
3	22/09/2006	11:10	22/09/2006	11:50	0:20
4	13/07/2008	6:40	13/07/2008	14:00	7:20
5	6/8/2008	12:30	13/8/2008	9:10	164:40:00
6	5/6/2009	7:50	5/6/2009	10:50	3:00
7	9/6/2009	7:45	9/6/2009	10:45	3:00
8	2/7/2009	11:05	2/7/2009	11:55	0:50
	Total outage hours				267:30:00

SEALING D/T OF DOWN STREAM

S.No.	Duration				Outage Hours
	From		To		
	Date	Time in Hours	Date	Time in Hours	
1	30/09/2006	9:05	30/09/2006	11:10	2:05
2	31/10/2006	10:00	31/10/2006	12:00	2:00
	Total outage hours				4:05

INTAKE GATE FAILURE

S.No.	Duration				Outage Hours
	From		To		
	Date	Time in Hours	Date	Time in Hours	
1	30/09/2006	9:05	30/09/2006	11:10	2:05
2	31/10/2006	10:00	31/10/2006	12:00	2:00
	Total outage hours				4:05

PILOT VALVE FAILURE

S.No.	Duration				Outage Hours
	From		To		
	Date	Time in Hours	Date	Time in Hours	
1	8/5/2008	16:15	8/5/2008	18:45	2:30
2	23/09/2009	23:05	24/09/2009	0:30	25:25:00
			Total outage hours		27:55:00

IRREGULAR VOICE OF MACHINE

S.No.	Duration				Outage Hours
	From		To		
	Date	Time in Hours	Date	Time in Hours	
1	31/01/2008	8:35	1/2/2008	16:30	31:55:00
			Total outage hours		31:55:00

THRUST BEARING FAILURE

S.No.	Duration				Outage Hours
	From		To		
	Date	Time in Hours	Date	Time in Hours	
1	11/8/2008	10:15	11/8/2008	15:30	5:15:00
2	15/09/2009	12:15	15/09/2009	12:50	0:35
			Total outage hours		5:50

UGB FAILURE

S.No.	Duration				Outage Hours
	From		To		
	Date	Time in Hours	Date	Time in Hours	
1	26/02/2009	9:45	2/3/2009	20:15	106:35:00
			Total outage hours		106:35:00

PROBLEM ON PMG

S.No.	Duration				Outage Hours
	From		To		
	Date	Time in Hours	Date	Time in Hours	
1	20/04/2009	8:20	20/04/2009	10:15	1:55
			Total outage hours		1:55

TGB FAILURE

S.No.	Duration				Outage Hours
	From		To		
	Date	Time in Hours	Date	Time in Hours	
1	26/02/2009	9:45	2/3/2009	20:15	106:35:00
			Total outage hours		106:35:00

APPENDIX 2.3
BREAK DOWN RECORD OF PATHRI HYDRO POWER STATION (PHPS)
Unit-3

PMG FAILURE

S.No.	Duration				Outage Hours
	From		To		
	Date	Time in Hours	Date	Time in Hours	
1	20/08/2005	7:30	20/08/2005	19:00	11:30
2	23/08/2005	9:00	23/08/2005	18:55	9:55
3	3/9/2005	8:55	3/9/2005	20:50	11:55
4	4/9/2005	8:10	4/9/2005	22:30	14:20
5	20/11/2008	8:15	20/11/2008	20:45	12:30
6	17/12/2008	9:30	17/12/2008	11:40	2:10
7	11/4/2009	5:30	11/4/2009	6:45	1:15
8	13/04/2009	17:30	13/04/2009	20:20	2:50
9	23/04/2009	19:45	23/04/2009	21:00	1:15
10	1/5/2009	10:15	1/5/2009	16:45	6:30
11	23/06/2009	9:05	23/06/2009	11:30	2:25
12	28/09/2009	9:15	28/09/2009	11:10	1:55
			Total outage hours		78:30:00

PILOT VALVE FAILURE

S.No.	Duration				Outage Hours
	From		To		
	Date	Time in Hours	Date	Time in Hours	
1	1/10/2007	11:40	1/10/2007	13:10	1:30
2	6/9/2008	12:30	6/9/2008	13:30	1:00
3	28/11/2008	6:00	28/11/2008	9:05	3:05
4	31/01/2009	17:30	31/01/2009	18:50	1:20
5	26/08/2009	0:10	26/08/2009	3:00	2:50
6	5/9/2009	17:40	5/9/2009	19:30	1:50
			Total outage hours		11:35

PROBLEM ON CARBON BRUSH OF EXCITER

S.No.	Duration				Outage Hours
	From		To		
	Date	Time in Hours	Date	Time in Hours	
1	7/8/2006	11:30	7/8/2006	11:55	0:25
2	22/06/2009	14:15	22/06/2009	18:05	3:50
	Total outage hours				4:15

FAILURE OF MACHINE DUE TO BRAKE PAD

S.No.	Duration				Outage Hours
	From		To		
	Date	Time in Hours	Date	Time in Hours	
1	7/3/2007	16:15	7/3/2007	18:15	2:00
2	7/8/2008	9:45	8/8/2008	17:30	31:45:00
	Total outage hours				31:47:00

FAILURE OF STATOR COOLER OF MACHINE

S.No.	Duration				Outage Hours
	From		To		
	Date	Time in Hours	Date	Time in Hours	
1	1/8/2007	10:05	1/8/2007	13:10	3:05
	Total outage hours				3:05

GOVERNOR PUMP FAILURE

S.No.	Duration				Outage Hours
	From		To		
	Date	Time in Hours	Date	Time in Hours	
1	9/8/2007	14:00	9/8/2007	14:30	0:30
2	1/10/2007	11:40	1/10/2007	13:10	1:30
	Total outage hours				2:00

COOLING PUMP FAILURE

S.No.	Duration				Outage Hours
	From		To		
	Date	Time in Hours	Date	Time in Hours	
1	13/04/2005	16:00	13/04/2005	17:00	1:00
	Total outage hours				1:00

INTAKE GATE GUIDE BEARING FAILURE

S.No.	Duration				Outage Hours
	From		To		
	Date	Time in Hours	Date	Time in Hours	
1	2/8/2005	9:05	2/8/2008	10:00	0:55
	Total outage hours				0:55

AUXILIARIES ROTOR & STATOR FAILURE

S.No.	Duration				Outage Hours
	From		To		
	Date	Time in Hours	Date	Time in Hours	
1	1/6/2007	7:05	3/6/2007	3:30	44:25:00
2	3/11/2009	10:00	7/11/2009	18:45	104:45:00
	Total outage hours				149:15:00

UGB FAILURE

S.No.	Duration				Outage Hours
	From		To		
	Date	Time in Hours	Date	Time in Hours	
1	15/07/2008	10:30	15/07/2008	15:45	5:15
2	7/8/2008	9:45	8/8/2008	17:30	31:45:00
3	14/02/2010	1:05	14/02/2010	4:05	3:00
	Total outage hours				40:00:00

THRUST COOLER FAILURE

S.No.	Duration				Outage Hours
	From		To		
	Date	Time in Hours	Date	Time in Hours	
1	30/07/2008	11:45	30/07/2008	14:05	2:20
2	6/9/2008	12:30	6/9/2008	13:30	1:00
	Total outage hours				3:20

FAILURE OF MACHINE DUE TO LOAD DECREASING

S.No.	Duration				Outage Hours
	From		To		
	Date	Time in Hours	Date	Time in Hours	
1	10/6/2007	15:05	10/6/2007	15:20	0:15
	Total outage hours				0:15

REFERENCES

1. IEEE Task Group, "A Four-State Model for Estimation of Outage Risk for Units in Peaking Service," *IEEE Transactions, vol. PAS-91*, pp. 132, April 1972.

2. Billinton, R., and Allan, R.N.: 'Reliability evaluation of engineering systems: concepts and techniques' (Plenum Publishing, New York, 1983)

3. Billinton, R., and Allan, R.N.: 'Reliability evaluation of power systems' (Plenum Publishing, New York, 1984)

4. Cunha, S.H.F., Gomes, F.B.M., Oliveira, G.L., and Pereira, M.V.F.: 'Reliability evaluation in hydrothermal generating systems', *IEEE Trans.*, 1982, PAS-101, pp. 4665-4673.

5. J. Endrenyi, "Three-State Models in Power System Reliability Evaluations", *IEEE Trans. Power Apparatus Syst, vol. 90,* July/Aug. 1971, pp. 1909-1916.

6. M.B. Guertin and Y. Lamarre, "Reliability Analysis of Substations with Automatic Modeling of Switching Operations", *IEEE Trans. Power Appuratus Syst,* vol. 94, no. 5, Sept./Oct. 1975, pp.1599 -1607.

7. R.N. Allan, R. Billinton and M.F. DeOliveira, "Reliability Evaluation of the Auxiliary Electrical System of Power Stations", *IEEE Trans.Power Appnrutus Sysr,* vol. 96, no. 5, Sept./Oct. 1977, pp.1441-1449.

8. R.N. Allan and J.R. Ochoa, "Modeling and Assessment of Station Originated Outage for Composite Systems Reliability Evaluation", *IEEE Trans. PowerSystenz,* vol. *3,* no. I, February 1988, pp.158-165.

9. R. Billinton, Hua Chen and Jiaqi Zhou, Generalized n+2 State System Markov Model for Station-Oriented Reliability Evaluation".IEEE PE-242-PWRS-0-01- 1997.

10. R. Billinton, Hila Chen, "Weaknesses of the Conventional Three State Model in Station-Oriented Reliability Evaluation", Microelectronic and Reliability, .accepted for publication,

11. R. J. Ringlee and S. D. Goode, "On Procedures for Reliability Evaluations of Transmission Systems," IEEE Trans. Power Apparatus and Systems, vol. 89, pp. 527-537, April 1970.

12. S. A. Mallard and V. C. Thomas, "A Method for Calculating Transmission System Reliability," IEEE Trans. Power Apparatus and Systems, vol. 87, pp. 824-834, March 1968.

13. F. E. Montmeat, A. D. Patton, J. Zemkoski and D. J. Cumming,"Power System Reliability, II - Applications and a Computer Program," IEEE Trans. Power Apparatus and Systems, vol. 84, pp. 636-643, July 1965.

14. A.R. Majeed and N.M. Sadiq, "Availability & Reliability Evaluation of Dokan hydro power station" IEEE Conf. Proc. Transmission and Distribution, 2006, pp. 1-6.

15. P. K. Varshney, A. R. Joshi, and P. L. Chang, "Reliability modeling and performance evaluation of variable link-capacity networks," IEEE Transactions on Reliability, vol. 43, 1994, pp. 378-382.

16. R. Billinton, Hau Chen, and Jiaqi Zhou, "Individual Generating Station Reliability Assessment", IEEE Trans. Power System, Vol. 14, No. 4, pp. 1238-1244, November 1999.

17. R.N. Allan, R. Billinton and M.F. DeOliveira, "Reliability Evaluation of the Auxiliary Electrical System of Power Stations", *IEEE Trans. Power Appnrutus Sysr,* vol. 96, no. 5, Sept./Oct. 1977, pp.1441-1449.

18. M. W. Gangel and R. J. Ringlee, "Distribution System Reliability Performance," IEEE Trans. Power Apparatus and Systems, vol. 87, pp. 1657-1665, July 1968.

19. R. Billinton and J. Tatla, "Composite Generation and Transmission System Adequacy Evaluation Including Protection System Failure Modes", *IEEE Trans. Power Apparatus Syst,* vol. 102, no. 6, June 1983, pp.1823-1830.

I want morebooks!

Buy your books fast and straightforward online - at one of world's fastest growing online book stores! Environmentally sound due to Print-on-Demand technologies.

Buy your books online at
www.morebooks.shop

Kaufen Sie Ihre Bücher schnell und unkompliziert online – auf einer der am schnellsten wachsenden Buchhandelsplattformen weltweit! Dank Print-On-Demand umwelt- und ressourcenschonend produziert.

Bücher schneller online kaufen
www.morebooks.shop

KS OmniScriptum Publishing
Brivibas gatve 197
LV-1039 Riga, Latvia
Telefax: +371 686 204 55

info@omniscriptum.com
www.omniscriptum.com

OMNIScriptum